Maha Elbana
Youssef Brouziyne
Editors

Irrigation Management

Strategies, Sustainability and Agricultural Advancements

nova science publishers

www.novapublishers.com

DOI: https://doi.org/10.52305/UQGT5734

NOTICE TO THE READER

The Publisher has taken reasonable care in the preparation of this book, but makes no expressed or implied warranty of any kind and assumes no responsibility for any errors or omissions. No liability is assumed for incidental or consequential damages in connection with or arising out of information contained in this book. The Publisher shall not be liable for any special, consequential, or exemplary damages resulting, in whole or in part, from the readers' use of, or reliance upon, this material. Any parts of this book based on government reports are so indicated and copyright is claimed for those parts to the extent applicable to compilations of such works.

Independent verification should be sought for any data, advice or recommendations contained in this book. In addition, no responsibility is assumed by the Publisher for any injury and/or damage to persons or property arising from any methods, products, instructions, ideas or otherwise contained in this publication.

The Publisher assumes no responsibility for any statements of fact or opinion expressed in the published contents.

This publication is designed to provide accurate and authoritative information with regard to the subject matter covered herein. It is sold with the clear understanding that the Publisher is not engaged in rendering legal or any other professional services. If legal or any other expert assistance is required, the services of a competent person should be sought. FROM A DECLARATION OF PARTICIPANTS JOINTLY ADOPTED BY A COMMITTEE OF THE AMERICAN BAR ASSOCIATION AND A COMMITTEE OF PUBLISHERS.

Additional color graphics may be available in the e-book version of this book.

Library of Congress Cataloging-in-Publication Data

ISBN: 979-8-89530-318-4 (Hardcover)
ISBN: 979-8-89530-532-4 (eBook)

Published by Nova Science Publishers, Inc. † New York

Dedication

To the research community, whose tireless efforts and discoveries illuminate the path of science, enhancing human well-being. And to everyone who deeply cares about our environment, and the preservation and sustainable management of our precious natural resources. Your dedication inspires and drives us towards a better future.

Contents

Acknowledgments

The editors extend their heartfelt appreciation to the collaborative authors of each chapter for their exceptional expertise and commitment. We are equally grateful to the diligent reviewers and Nova Science Publishers, Inc. for their meticulous efforts, insightful feedback, and dedication to presenting this book in an exemplary format for the agricultural community and environmental advocates.

Prologue

Water is the essence of life and the cornerstone of agriculture. Yet, in the 21st century, this precious resource faces unprecedented pressures. Climate change, population growth, urban expansion, and soil degradation have converged to create a complex web of challenges. Agriculture, responsible for ~70% of global water consumption, stands at the heart of this crisis. Yet, agriculture is a key player in the world's food production and security. Researchers, policymakers, and farmers are facing a scary question: how the growing population can be fed without deteriorating and threatening our limited water resources?

Irrigation Management: Strategies, Sustainability, and Agricultural Advancements, invites readers on an enlightening journey through the dynamic world of modern irrigation. At its core, it explores how innovative sustainable natural resources management, and a deep understanding of environmental systems can reform agriculture for the better. From climate-resilient irrigation techniques to artificial intelligence-powered water management systems, the chapters delve into groundbreaking strategies designed to improve soil properties to optimize crop and water productivity.

Real-world examples highlight how the communities are adapting to the challenging change of climate. The text bridges disciplines, bringing together insights from agronomy, environmental science, engineering, and policy to present a comprehensive roadmap for the future of irrigation.

Above all, this book celebrates the human spirit of resilience and creativity. It honors the investigators developing water-saving technologies, the farmers embracing climate-smart methods, and the policymakers encouraging sustainable water management. In doing so, it offers not only hope but a comprehensive scheme for a more sustainable, productive, and fair agricultural future.

Let this prologue serve as an invitation - a call to engage with the ideas and solutions that can ensure a water-secure world. As the chapters ahead demonstrate, the challenges are immense, but so too are the opportunities to innovate, adapt, and thrive.

Chapter 1

Climate-Resilient Irrigation: Adapting to a Changing World

Jitendra Singh[1]
Atin Kumar[2,*]
Nitish Karn[2]
Rohit Kumar[3]
Amir Parnian[4]
Mina Mahbod[5]
and Mohamed A. E. AbdelRahman[6]

[1]Assistant Professor-cum-Jr. Scientist,
Department of Soil Science and Agricultural Chemistry,
Veer Kunwar Singh College of Agriculture Dumraon,
Bihar Agriculture University, Sabour, Bihar, India
[2]School of Agriculture, Uttaranchal University, Dehradun, Uttarakhand, India
[3]Faculty of Agricultural Sciences, GLA University, Mathura, Uttar Pradesh, India
[4]National Salinity Research Center (NSRC),
Agricultural Research Education and Extension Organization (AREEO), Yazd, Iran
[5]Department of Educational Sciences, Farhangian University, Tehran, Iran
[6]Division of Environmental Studies and Land Use,
National Authority for Remote Sensing and Space Sciences (NARSS), Cairo, Egypt

Abstract

The chapter "Climate-Resilient Irrigation: Adapting to a Changing World" explores the imperative of climate-resilient practices in the field

* Corresponding Author's Email: atinchaudhary0019@gmail.com

In: Irrigation Management
Editors: Maha Elbana and Youssef Brouziyne
ISBN: 979-8-89530-318-4

of irrigation to address the challenges posed by a changing climate. As global temperatures rise and weather patterns become increasingly unpredictable, it is vital to adapt our agricultural practices to ensure food security and environmental sustainability. The first half of this chapter gives a thorough overview of how climate change affects agriculture. It highlights the pressing need for climate-resilient irrigation techniques. In contrast, the second half of the chapter explores various irrigation practices and technologies designed to be resilient to climate change, aiming to improve water efficiency and counteract its adverse effects. The strategies encompass a variety of irrigation methods, such as drip and sprinkler irrigation, as well as water management approaches like rainwater harvesting and recycling, alongside the adoption of crop varieties suited to changing environmental conditions. The importance of maintaining soil health and employing conservation techniques, such as no-till farming and cover cropping, is also discussed. Technological innovations play a crucial role in climate-resilient irrigation, and the chapter highlights the applications of remote sensing, GIS, weather forecasting, and smart irrigation controllers. Automation and robotics are presented as pivotal tools in optimizing irrigation processes.

Effective policy and governance are central to the success of climate-resilient irrigation. The chapter examines government initiatives, water resource management, legal frameworks, and international collaborations that support these practices. Real-world case studies display successful implementations of climate-resilient irrigation in various contexts, from arid regions to coastal areas and small-scale community initiatives to large commercial farming operations.

Challenges and future directions in climate-resilient irrigation are explored, including economic and technical barriers, socio-economic impacts, and the need for research and development. The chapter wraps up by highlighting the main points and proposing future steps to develop climate-resilient agriculture, promote the sustainable utilization of water resources, and safeguard global food production amidst the challenges posed by climate change.

Keywords: climate-resilient irrigation, water efficiency, climate change adaptation, irrigation technology, sustainable agriculture, water resource

List of Abbreviations

A&FE	Agriculture and farmers' empowerment
AI	Artificial Intelligence
ANFIS	Adaptive neuro-fuzzy inference systems

ANN	Artificial neural networks
BTC	Classifiers for boosted trees
CO_2	Carbon dioxide
CSA	Climate smart agriculture
DANN	Dynamic artificial neural network
DL	Deep learning
DNN	Deep neural networks
DoWR	Department of water resources
ELM	Extreme learning machine
F&ARD	Fisheries and animal resources development
FTDNN	Focused time-daily neural network
GBRT	Gradient boosted regression trees
GDP	Gross domestic product
GPM	Gallons per minute
IoT	Internet of things
IWRM	Integrated water resources management
KNN	K-nearest neighbours
ML	Machine learning
MoWR, RD & GR	Ministry of Water Resources, River Development, Ganga Rejuvenation
OIIPCRA	Odisha Integrated Irrigation Project for Climate Resilient Agriculture
PMKSY	Pradhan Mantri Krishi Sinchayee Yojana initiative
PPs	Participatory planning
PVC	Polyvinyl chloride
RWC	Relative leaf water content
SVM	Support vector machines.

Introduction

The effects of climate change present distinct challenges for the agriculture sector. Climate change affects groundwater supply and distribution, snowmelt, river flow, precipitation availability, and improved water quality. Water resources and their importance for attaining sustainable development are closely related to climate change adaptation. Numerous adjustments are necessary to address climate change effectively, improve existing land and water management capabilities, bolster resilience against climate impacts, and strengthen water security (Devi et al., 2023). 70% of the water in the world is used for agriculture. According to FAO, this percentage will keep rising to satisfy the expanding needs of the population (FAO, 2021). Therefore, the task at hand is to effectively utilize water resources in agriculture to sustainably provide for the expanding global population and ensure food security. Climate change impacts all types of water, including rain-fed water and irrigated water used for agriculture, animal feed, and food production. Plant growth and production is dependent on water availability. Rainfall is the natural source of water for plants. When there is not enough rain, plants need more water. This is where irrigation comes in, as it ensures a controlled and steady water supply for crops to obtain the maximum yield from crops.

In agriculture, irrigation is the artificial manipulation of groundwater to give plants the moisture they require to thrive. According to the International Commission on Irrigation & Drainage, worldwide irrigated areas for developed, developing and least developing countries are 51.761 million ha, 245.837 million ha and 10.321 million ha, respectively (Figure 1). Water resource management is the process of strategically planning, designing, allocating and managing water resources to ensure they are used efficiently. Water scarcity is a major cause of famine and drought (Chamara et al., 2022). Climate change is poised to reshape the global agricultural landscape, potentially offering opportunities for rural communities in harsh environments reliant on farming. However, farmers face direct threats such as crop failure, livestock loss, and soil erosion. Climatic factors have a major role in affecting agricultural activities, from land preparation to harvesting and storing, leading to the need for farmers to understand climatic factors and their interaction with crops. Modifying the balance and concentration of greenhouse gases in the atmosphere, including water vapor, methane, carbon dioxide, ozone, etc., is accountable for climate change. The rise in greenhouse gas emissions negatively impacts nutrition, water security, and other social upheavals. Atmospheric pressure gradually intensifies the greenhouse effect, causing a steady rise in the global average temperature. The concentration of carbon dioxide (CO_2) has already surpassed 400 ppm (Mikhaylov et al., 2020). Starting in 1880, the temperature of the Earth has been increasing at a rate of 0.08°C per decade. This rate has more than doubled in the last 40 years, increasing to 0.18°C per decade since 1981. As of 2020, surface temperatures were 0.98°C higher than the average temperature of the twentieth century, which was 13.9°C. Additionally, temperatures were 1.19°C higher than during the pre-industrial period of 1880-1900 (Sanchez-Lugo et al., 2019). River delta expansion faces risks from subsidence, altered flood patterns, salinity intrusion, and rising sea levels. Changes in water availability, including earlier glacier runoff and increased crop water demand due to higher temperatures, pose additional challenges (Chowdary et al., 2023). Effective water resource management, infrastructure development, rainwater harvesting, and community-based land and water management initiatives are essential. Agricultural water management is critical due to overpopulation and water scarcity, driving interest in climate-smart agricultural practices (CSA) that consider social, economic, and environmental factors. Despite the longstanding use of traditional irrigation systems, their effectiveness is often limited. Smart irrigation systems leverage advanced technology to provide a more efficient and effective method for cultivating crops. Water, crucial for crop growth, necessitates careful quality management to ensure optimal yields. With the rise of the global population, the imperative to produce more food with diminishing water resources becomes increasingly pressing (Bezas et al., 2023).

In agriculture, timing and proper water allocation in irrigation is crucial. Farmers rely on weather forecasts to inform their irrigation practices. By using real-time temperature and soil moisture monitoring in conjunction with meteorological forecasts, intelligent autonomous irrigation systems can forecast the drought occurrence. Prioritizing water conservation is as important as determining when to turn on irrigation systems (Feng et al., 2017).

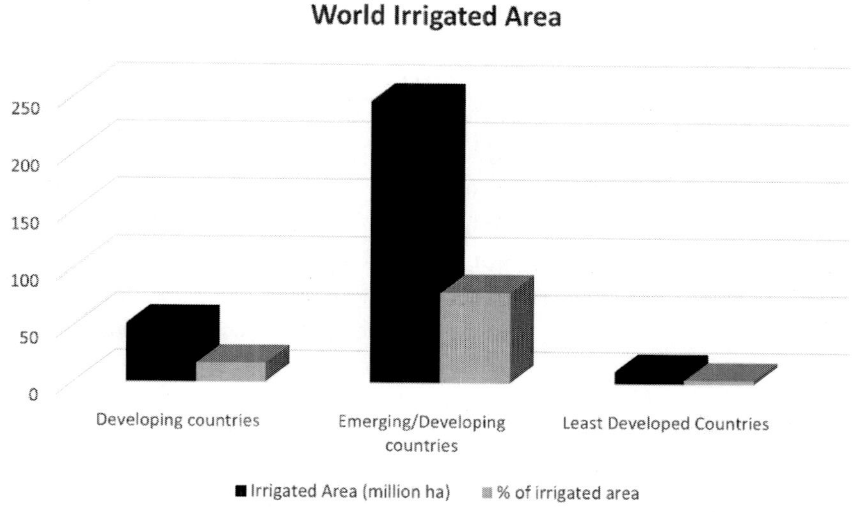

Figure 1. World Irrigated area (ICID, 2018).

Impact of Climate Change on Agriculture

India ranked 4th on the list of countries most affected by climate change in 2015 (Ministry of Earth Sciences, GOI, 2020). Climate change is negatively impacting agriculture, which is particularly vulnerable due to its dependence on monsoon rainfall and the engagement of a large population in agriculture. Decreases in daily temperature range in many areas, with night minimum temperature increasing more than diurnal maximum temperature, increases in the frequency, magnitude, and duration of climate-related extremes, such as heat waves, intense rainfall, tropical storms, cyclones and wildfires, floods, as well as an increase of regions affected by droughts, are already occurring (Wassmann et al., 2010; IPCC, 2013).

It has majorly affected crop yields, water resources, pests and diseases, and overall food security. Environmental elements, like weather conditions and soil quality, influence plant growth and development. The weather conditions during the harvest period significantly impact the outcome of the crop, determining its success or failure. While addressing climate change presents challenges and requires investment, adjusting cropping patterns and adopting suitable agricultural practices can enhance crop yields by mitigating the impacts of unfavorable weather conditions. The main factor causing climate change is an increase in the emissions of greenhouse gases such carbon dioxide, methane, nitrous oxide, and fluorinated gases (Islam et al., 2021). Floods and droughts are brought on by these gases, which also alter temperature and precipitation patterns, thereby lowering the quality of water and soil. According to Richard et al., 1998 climate change causes a shift in runoff and groundwater recharge rates, which disturb water supplies and cause changes in capital or technological requirements such as irrigation methods and surface water storage. Unfortunately, the pace of current climate change is alarming, primarily attributed to human activities such as deforestation and the extensive use of fossil fuels for transportation, industrial operations, and residential needs. Climatic factors affect all stages of agricultural activity, from tillage to storage, emphasizing the need for farmers to have a sound understanding of these factors and their interaction with crops (Sadras et al., 2016). Some significant impacts of climate change on agriculture are mentioned.

Erratic Monsoon Patterns

Monsoons are essential for agriculture because they provide water for cultivation and determine planting and harvesting processes. Significant modifications to monsoon patterns brought about by climate change include adjustments to the timing, duration, and intensity of rainfall. Inadequate and erratic monsoons can lead to droughts or excessive rainfall, leading to water shortages or intrusions, which can affect crop production (Lu et al., 2005). Rainfall in northern India is governed by the southwest monsoon, which downpours from the middle of June till the end of September. Temporal delay, variability and change in spatial rainfall patterns are of great concern in tropical arid and semi-arid zones, since - in those regions - natural vegetation and agricultural ecosystems are highly sensitive to slight variations in rainfall (Singh et al., 1992). Weak monsoons in successive years affect agricultural activities and aquifer recharge. The western part of India often suffers from drought due to poor rainfall and high heat flow, and the overexploitation of limited groundwater resources makes the situation worse (Bhuiyan et al., 2009).

Water Stress and Irrigation Challenges

Climate change affects the availability of water needed for agriculture. Changing rainfall patterns and increased evaporation due to rising temperatures can lead to water stress and scarcity. Changes in precipitation patterns including droughts or excessive rainfall can result in water scarcity or waterlogging, affecting irrigation practices. Limited water availability can reduce crop water uptake, leading to water stress and decreased yields.

Heat Stress on Crops

Plants can become stressed when exposed to high temperatures for long periods. If farmers do not take precautions to prevent heat, heat stress can affect plants and reduce yield. However, the symptoms of heat stress in plants are not always obvious, as they are similar to other diseases. To know if crops are under stress and protect them in a timely manner, it is essential to use accurate weather data and other advice, such as that provided by those at the highest levels of agriculture. The escalation of temperatures due to climate change will heighten the vulnerability of crops prone to drought. Elevated heat levels during critical growth stages can harm crops, diminish photosynthesis, and disrupt fruiting processes, leading to decreased yields. Heat stress is a major problem for many crops, such as vegetables, rice, wheat, and maize. Heat stress in crops manifests itself in many ways. However, it can be challenging to identify because its symptoms are similar to those caused by some pests and diseases. Additionally, cold weather can cause plant diseases. Some crucial symptoms caused by heat stress in crops are as follows:

Heat stress can cause dehydration in plants, inhibiting their growth. Reduced photosynthetic efficiency directly results from the effects of stress on relative leaf water content (RWC) and plant water. Therefore, heat stress can cause plants to suffocate and wilt. However, plants can control air and airflow to maintain thermal balance under temporary or moderate heat stress conditions. Due to high temperatures, protein and sugar concentrations can be adjusted to maintain the osmotic pressure in the plant. Finally, crop losses are witnessed due to heat stress, affecting everything from rice and legumes to root vegetables.

Increased Frequency of Extreme Weather Events

Climate change will increase the frequency and power of extreme weather events such as cyclones, typhoons, floods, and droughts. These events pose significant risks, resulting in crop damage, soil erosion, livestock loss, and infrastructure destruction. Farmers encounter difficulties in recovering from such adversities and restoring their means of livelihood. Experts quickly attribute previous years' extreme monsoon rains and heat waves to possible El Niño, predicting droughts, average monsoon rains and warmer conditions for this year. According to Swaminathan and Rengalakshmi, 2016, the growing incidence and severity of droughts, floods, hailstorms and other extreme weather events severely affect the livelihood options for small-scale farmers. It is disturbing to note that India loses, annually, about 2% of its GDP (Gross Domestic Product) and around 12 percent of central government revenues to natural disasters (Lester and Gurenko, 2003). Tropical storm and hurricane frequencies vary considerably yearly, but evidence suggests substantial increases in intensity and duration since the 1970s (Meehl et al., 2007). The study conducted by Hirabayashi et al., (2008) projected a possibility of hydrological drought and an increase in the risk of floods in humid monsoon regions of South Asia. Similarly, Rajendran and Kitoh (2008) and Orlowsky and Seneviratne (2011) indicated the chances of an increase in extremely hot days over East Asia and South Asia.

Shifting Crop Suitability

Climate change can alter the suitability of specific crops in different parts of the world. Fluctuations in temperature and precipitation can render traditional crops unsuitable in some areas, while creating opportunities for different crops in others. According to Singh, I. K., 2023, India is experiencing shift of seasons due to climate change. The impact of this climate change has been felt since the beginning of Rabi rice harvest in Punjab and Haryana last year. Heatwaves and untimely rains affected last year's rice harvest and caused damage in central India during the festive season. The early onset of summer was reported in Lahaul, Kashmir, and other parts of the country. Rain patterns changed over Meghalaya, India's wettest state, causing losses to mountain agrarian communities. In southern India, down from Telangana, Rabi was delayed due to a more extended rainy spell. Many farmers across the country lost out and even had to re-sow, due to excessive moisture in their fields. In northern India now, mustard is harvested in February, earlier; it usually happens around and after Holi. However, high temperature causes the mustard crop to ripen early. Many regions of Rajasthan, which harvested mustard in early and mid-February, also complained about the decrease in production. Consequently, farmers must adapt by altering their crop selections and embracing new varieties or agricultural techniques (Zou et al., 2012).

Increased Pest and Disease Pressure

Climate change can affect population dynamics and the geographic distribution of pests and diseases. Elevated temperatures promote the growth of pests, leading to increased crop infestations. Alterations in rainfall patterns also create conducive conditions for spreading diseases, resulting in crop damage and reduced yields.

Climate- Resilient Irrigation

Increasing temperatures due to climate change have increased the possibility and intensity of drought and monsoon disturbance. So, appropriate water management would increase the resilience of agricultural production to climate change. Due to the constraint of water resources, the supply of agricultural irrigation water cannot be expanded blindly. Many agricultural practices can mitigate the effect of climate change on agriculture and manage water efficiently for agriculture. Some of the practices are:

Sprinkler Irrigation

This method involves spraying water over the crops using pipes and sprinkler heads. It is a very resourceful and can be used for a wide range of crops, but it requires high costs and regular maintenance.

Sprinkler irrigation is a technique that consists of discharging water into the air, letting it descend onto the soil surface, resembling natural precipitation (Table 1). Water under pressure is channeled through tiny nozzles or orifices, often using a pump, creating spray. By carefully choosing the appropriate nozzle sizes, operational pressure, and sprinkler spacing, it is feasible to achieve a nearly even distribution of irrigation water and replenish the crop root zone at a rate that aligns with the soil's infiltration rate (Kumar et al., 2024).

Various Components of a sprinkler irrigation system (Figure 2):

1. A pumping unit
2. Debris removal equipment
3. Pressure gauge
4. Pipelines
5. Couplers
6. Riser pipes
7. Sprinklers
8. Additional components commonly used in irrigation systems include valves, elbows, tees, couplings, plugs, and other fittings.

Table 1. Sprinkler irrigation system (Vijai et al., 2023)

S. No.	Type of Sprinkler	Precipitation Rate (mm h^{-1})
1	Low Volume	<13
2	Medium Volume	13-25
3	Large Volume (Rain gun)	>25

Pumping Unit

This is an integral part of the system and is responsible for taking water from a water source, such as a well, pond or reservoir, and sending it to the sprinklers. The most commonly used pump types for this purpose are high-speed centrifugal pumps or turbo-pump driven by an electric motor or a diesel engine. The suction part of the pump is equipped with a valve to the bottom to prevent water from flowing back, and the outlet pipe is connected to the main pipeline of the system.

Pipeline

A sprinkler system has two types of pipelines: mains and laterals. The primary channel transports water from the pumping unit to various field areas, while secondary pipelines distribute water from the main conduit to the lateral lines. Meanwhile, the lateral pipeline distributes water from the primary or secondary pipeline to the sprinklers.

Couplers

Couplers connect two pieces of tubing or to connect tubing to fittings in a sprinkler system.

Sprinklers

Two types of sprinklers are available; rotating and fixed types. Rotating sprinklers are diverse and can be adjusted to accommodate various application rates and spacing needs. They work well within the 10 to 70-meter head pressure range at the sprinkler, but 16 and 40-meter head pressure is optimal for most fields. On the other hand, stationary sprinklers are typically employed for watering smaller spaces like lawns and gardens.

Water Meters

These instruments gauge the quantity of water the sprinkler system dispenses.

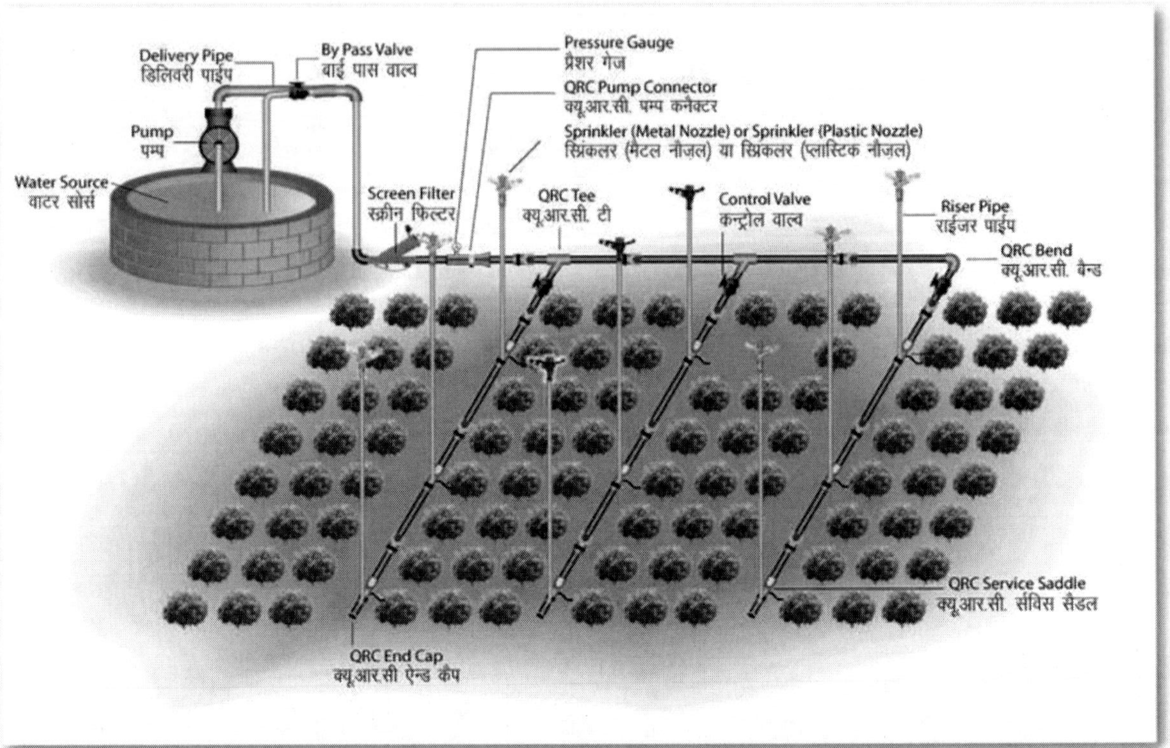

Layout of Sprinkler Irrigation System (छिड़काव सिंचाई प्रणाली का रेखाचित्र)

Figure 2. Layout of Sprinkler Irrigation System.

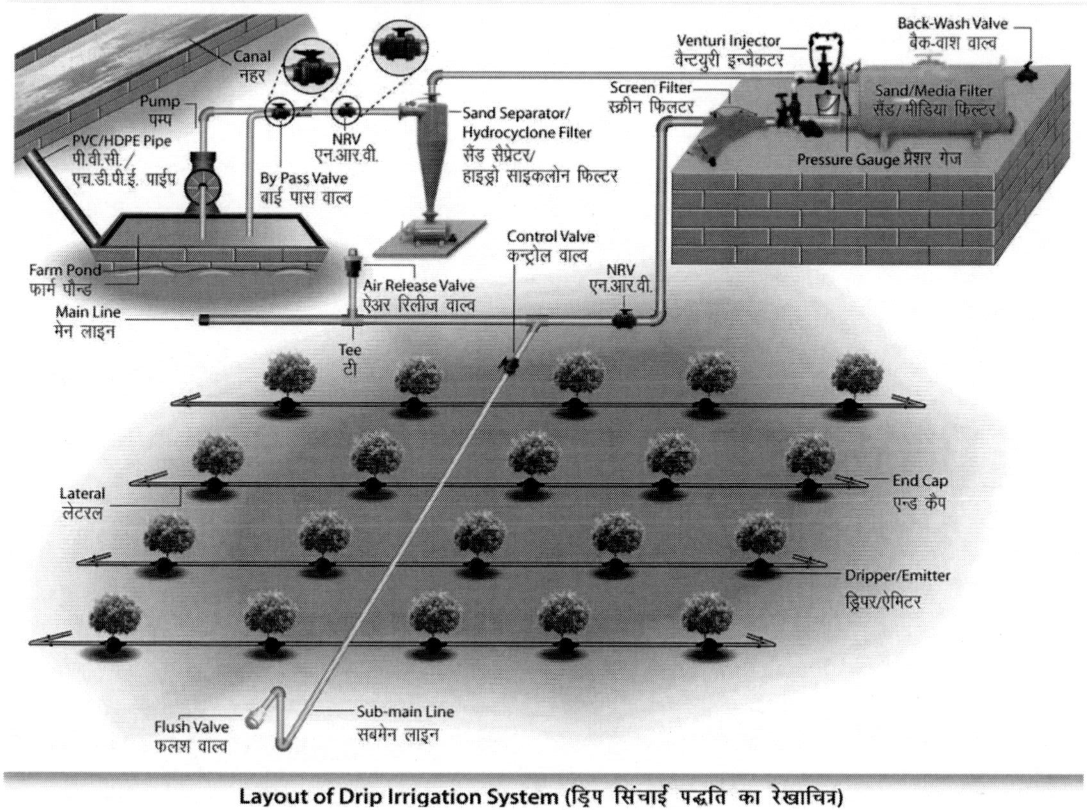

Figure 3. Layout of Drip Irrigation System.

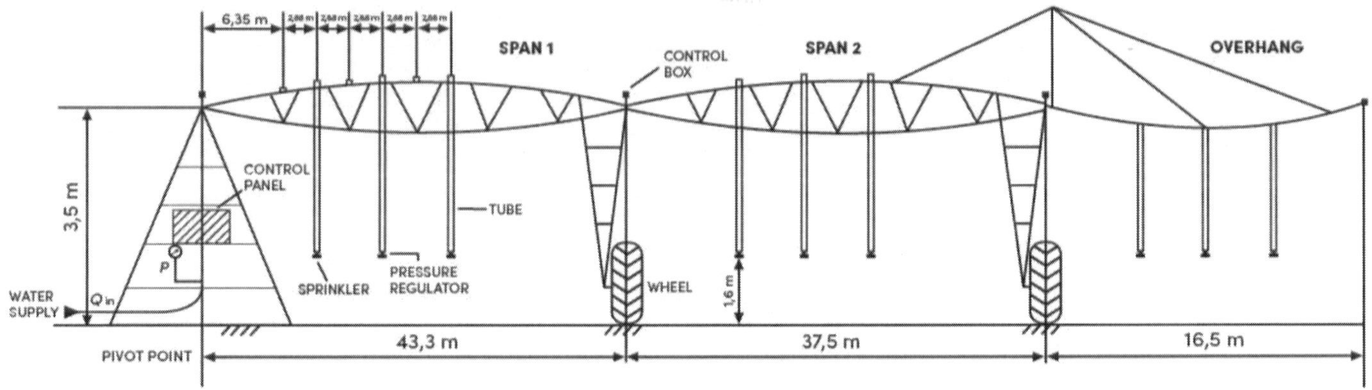

Figure 4. Diagram of a Center Pivot Irrigation System (Agrivi, 2023).

Pressure Gauges

They are essential in maintaining the desired pressure for uniform water delivery through the sprinkler system.

Debris Removal Equipment

When utilizing surface water sources like streams, ponds, or canals, debris removal equipment is required. It prevents debris like sand, leaves, and sticks from plugging the sprinklers.

Drip Irrigation

Trickle irrigation, another name for drip irrigation, is a type of micro-irrigation designed to gradually provide water to plant roots, whether the roots are buried beneath the soil or right on top of it. The primary objective is to target water delivery to the root zone while minimizing evaporation. Drip irrigation uses a low-pressure form of water delivery, unlike sprinkler systems. It is recognized for its high efficiency in crop irrigation and presents numerous benefits compared to conventional irrigation techniques, especially in reducing evaporation rates (Kumar et al., 2023). Drip irrigation involves using small, perforated pipes placed either underground or slightly above the soil surface adjacent to the crops, through which water drips gradually onto the plant roots and stems. Drip irrigation delivers precise amounts of water for irrigation and administers fertilizer tailored to the crop's requirements, thereby optimizing nitrogen use efficiency and fostering root growth. Drip irrigation enhances fertilizer uptake by delivering small amounts of fertilizer directly to the crop's root zone, thereby maximizing its utilization with less water and reducing soil salinity. This method, highly effective in dry regions, has been widely used to mitigate water scarcity because it efficiently uses water resources (Shivanna, 2022).

Drip Irrigation Components

The usual elements of a drip irrigation setup include a pump or raised tank, a main pipeline, sub-main lines, lateral pipes, as well as emitters (Figure 3). In a drip irrigation setup, water moves from the main pipeline to the secondary conduits and then to the lateral tubes. The laterals feature emitters that dispense water to the plants for irrigation. Typically, black PVC (polyvinyl chloride) tubing is used to manufacture the primary, secondary, and lateral lines, and emitters. Apart from these components, essential elements of a drip irrigation system encompass regulators, filters, valves, water meters, and devices for applying fertilizer.

Pump

To flow water through the pipelines - mainline, lateral lines, drippers, and emitters - the pump generates the necessary pressure. Pumping systems can be powered by electricity, diesel, gas, or solar energy. Since drip irrigation systems operate at low pressure, gravity can also pressurize the drip lines. Typically, centrifugal pumps driven by electric motors are the most commonly used. Laterals can function at low pressures ranging from 0.15 to 0.2 kg cm^{-2}, or at higher pressures between 1 to 1.75 kg cm^{-2}. The water released from the emitters typically maintains or approaches atmospheric pressure.

Tank

In certain drip irrigation setups, a container, commonly referred to as a chemical reservoir, is positioned at the start of the system. This reservoir dispenses herbicides, fertilizers, and other chemicals into the field in conjunction with the irrigation water.

Filter

The filtration system is crucial in drip irrigation systems because it prevents pipe and emitter/dripper blockages. Typically, this system comprises valves and a pressure gauge to manage and regulate the water flow.

Emitters

Drip nozzles, also known as emitters or drippers, are positioned alongside the lateral pipeline within a drip irrigation system. These emitters facilitate the gradual release of water, typically in small droplets. The size and pressure of the opening determine the volume of water discharged by each emitter over time. Typically, emitters release water at rates ranging between 2 - 10 liter hr^{-1}. Micro-tubes can also be incorporated into the lateral pipes of a drip irrigation system, serving multiple functions, including acting as emitters, connectors, and pressure regulators.

Central Pivot Irrigation System

Central pivot irrigation is a mechanized pressurized water irrigation system (Figure 4) that utilizes rotating sprinklers attached to a long horizontal arm. This arm is linked to a central pivot, typically a well or water source. While the pivot rotates, the sprinkler heads distribute water in a circular motion, moving around a central point within the field to water the crops below. This method effectively spans large areas with few operational requirements and is frequently used in areas with low precipitation and high evaporation rates (Shilpa et al., 2019). A pivot setup consists of a horizontal stainless steel pipeline placed on movable truss structures set up across neighboring segments. Each truss is fitted with motorized wheels, allowing the entire pipeline to move across the field. Sprinkler nozzles are attached along the pipeline or suspended from the lateral pipe using smaller pipes or cables. Flow rates, measured in gallons per minute (GPM), can be easily modified within a center pivot system, and pressure regulators can be utilized to set up either a high-flow or low-flow arrangement tailored to specific water consumption needs (Kumar et al., 2022).

Water Management

The agricultural sector heavily relies on water, making it crucial for its operations. Nevertheless, as water scarcity becomes more prevalent worldwide due to climate change, population expansion, urban development, industrial activities, and intensified agricultural methods, it is crucial to address this matter comprehensively (Vohra and Saxena, 2019). This entails water conservation, efficient resource utilization, and investment in new technologies and infrastructure to enhance water availability and quality. Effective water management is essential for maximizing agricultural production and minimizing water wastage. It encompasses strategic planning, development, distribution, and sustainable utilization of water resources in adherence to established laws and regulations. Implementing suitable agricultural approaches, crop rotation plans, and enhanced methods for water distribution, such as land grading and drip/sprinkler irrigation, can bolster water efficiency and adaptability to fluctuations in climate. Irrigation management is an area of extensive study, with innovative technologies such as machine learning models being utilized to optimize water usage (Yang et al., 2023). Techniques like dividing farm fields into different management zones and employing variable rate irrigation enable precise control over irrigation rates, contributing to efficient water utilization. As a result, by implementing effective water management techniques, it is feasible to control water quality, make the most use of available water, and ultimately increase crop yields (Patel et al., 2020).

There are some techniques by which water for can be judiciously managed.

Rain Water Harvesting

Insufficient water availability during the dry season poses a significant challenge for farmers aiming for successful cultivation. This issue can be mitigated through rainwater harvesting and its prudent utilization in agricultural practices. Rainwater harvesting involves gathering and storing rainwater using specially designed systems positioned in naturally occurring or human-made catchment areas. This practice can prove highly advantageous for farmers by facilitating the irrigation of crops during periods of moisture scarcity in the dry season (Senthil et al., 2023). There are three primary methods of rainwater harvesting:

- Capture of surface run-off: Water is directed through channels for later use.
- Recharging groundwater: A natural process where water seeps from the surface into underground aquifers.
- Use of sedimentation tanks for domestic or industrial purposes.

Water Recycling/Reuse

Water reuse involves reclaiming and treating impaired waters for beneficial purposes. This process primarily transforms non-potable wastewater into a valuable resource, thereby reducing the economic and environmental burdens of establishing new water sources. Water recycling and reuse encompass wastewater collection, treatment, and utilization, particularly from municipal, industrial, and agricultural sources (Kumar et al., 2023). If appropriately treated, the treated water can be used for irrigation, industrial activities, and potentially domestic use. In some instances, treated wastewater is indirectly utilized for drinking purposes, such as injecting it into groundwater aquifers to enhance capacity and mitigate saltwater intrusion. Utilizing water recycling and reuse strategies is essential for adapting to the challenges brought by climate change, such as erratic weather conditions, extreme droughts, and increasing sea levels, which endanger both the availability and purity of freshwater reserves (Mohammadi et al., 2015).

Water Management for Drought Prone Areas

Taking steps for mitigation and preparedness is crucial for minimizing the effects of drought, including actions like prediction, alert systems, drought insurance, and disaster assistance programs. In areas susceptible to drought, readiness should be incorporated into the management of water resources (Kumar et al., 2017). Drought mitigation efforts can be categorized into three main groups: increasing water supply through water-supply measures, reducing water demand during drought through water-demand measures, and implementing impact-minimization measures to lessen adverse effects. Irrigation emerges as the most efficient tactic for resilience against drought and holds a prominent position in maintaining stability in agricultural output. The government has taken several actions to alleviate the negative effects of drought, such as watershed management, cultivation techniques suitable for arid regions, soil moisture preservation, and the uptake of sprinkler and drip irrigation technologies (Sadras et al., 2016).

Machine Learning and Artificial Intelligence Techniques in Water Management

Water irrigation is a heavily researched field that extensively employs machine learning (ML) models to enhance water usage efficiency. These models segment farm fields into distinct management zones, allowing for precise control over irrigation rates through variable-rate irrigation techniques. Artificial intelligence (AI) and ML have emerged as key technologies in resolving water scarcity concerns by managing, forecasting, and optimizing water supplies. Forecasting water demand holds excellent potential for efficiently distributing the available water resources. Mohapatra et al., (2021) used Adaptive Neuro-Fuzzy Inference Systems (ANFIS), Deep Neural

Networks (DNN), and Support Vector Machines (SVM) to estimate seasonal groundwater levels. They evaluated these methods' effectiveness with appropriate statistical and graphical indicators. The DNN technique performed better than the other methods in accurately predicting seasonal groundwater levels.

In order to forecast groundwater quality, Singha et al., (2021) developed a deep learning (DL) model that was contrasted with three machine learning (ML) models: XG Boost, random forest, and artificial neural networks (ANN). Their findings demonstrated that the DL model exhibited superior predictive performance in terms of the performance metric R2. In order to estimate the irrigation water quality index, Yu et al., (2022) used four machine learning techniques: support vector machines (SVM), ANN, Extreme learning machine (ELM), and random forests. They used just the physical properties of groundwater as inputs, such as pH, dissolved oxygen, temperature, and electrical conductivity. The findings indicated that these models were successful in accurately estimating the irrigation water quality index. Ghiassi et al., (2017) used the k-nearest neighbors (KNN) method, a focused time-delay neural network (FTDNN), and a dynamic artificial neural network (DANN) to anticipate water consumption on a daily, weekly, and monthly basis. Three machine-learning techniques were used to create several irrigation suggestion models: first, the conventional linear regression technique; second, gradient boosted regression trees (GBRT), a nonparametric method; and third, classifiers for boosted trees (BTC) Models like GBRT and BTC needed fewer adjustments to account for non-linear connections between variables than traditional linear regression. The final model might greatly help agronomists in their planning of irrigation systems.

Irrigation Water Requirement Prediction Using ML

The application of machine learning in water management has been discussed in several studies that are summarized in (Table 2). Generally, implementing an irrigation prediction approach for intelligent automatic irrigation involves several steps:

- First, sensors such as soil moisture, temperature, and rain gauges are installed.
- These sensors are then connected to an acquisition system.
- The Node-RED platform is utilized to enable data storage, alert notifications, and monitoring.

- The gathered data is then processed by various methods, such as Naive Bayes, K-nearest neighbors (KNN), neural networks, support vector machines (SVM), and logistic regression. The KNN algorithm outperformed the other techniques, with a 98.3% decision accuracy.

Table 2. Applications of machine learning (ML) in the management of water

References	Application	Functionality/Feature	Algorithms/ML models
(Feng et al., 2017)	Estimation of Evapotranspiration in arid and semi-arid regions	Linked to a weather forecasting station and trained and evaluated using data that has been trained	ANN/GRN
(Islam et al., 2021)	Real-time smart irrigation system	Temperature, Soil moisture, and Humidity	DT
(Sun et al., 2021)	Estimation of Evapotranspiration in arid and semi-arid regions	Meteorological data, Temperature, Humidity, and Wind speed.	SVM-RBF/ MARS
(Goap et al., 2018)	Irrigation requirements predictions	UV Sensor, Temperature, Soil Moisture, and Humidity.	K-means SVR
(Dimitriadis and Goumopoulos, 2008)	Management of natural resources in farming land.	Agricultural data	OneR, ZeroR, NaïveBayes Simple, J48 DecisionStump
Sun et al., 2021)	Global hydrological model: Prediction and data gap filling	ERA5-Land (Precipitation, Temperature). Evapotranspiration; NDVI Vegetation Index	Automated ML: CNN; MLR; MLP; ARX–; SARIMAX–ARIM; XGBoost–eXtreme
(Mohammadi et al., 2015)	Daily dew point temperature.	Meteorological data (humidity, air temperature, solar radiation and vapor pressure)	ELM

Effective Policies and Governance for Climate-Resilient Irrigation

Irrigation systems play a vital role in the world's agriculture, receiving significant support from the public sector to foster long-term growth. State governments undertake the planning, funding, execution, and maintenance of water resource projects based on their respective resources and priorities. Various types of irrigation systems contribute to agricultural development, involving the water supply to fields either through gravity or pumping mechanisms. Water can be sourced from rivers, canals, lakes, reservoirs, groundwater, or retained directly at the cultivation site (Pathmudi et al., 2023). A successful irrigation system is defined by reliable access to sufficient water precisely when and where crops require it. The irrigation scheme should meet the following criteria:

1. Feasible hydraulically, considering factors such as raw water availability.
2. Properly designed, including considerations for storage capacity, conveyance capacities, and control structures.
3. Efficiently operated, ensuring effective water allocation within the scheme.

The scheme needs to be financially viable, indicating that the economic gains surpass the operational and maintenance expenses by a margin that appeals to farmers. Such a scheme typically exhibits good or reasonable climate resilience, as it is intended to withstand normal weather fluctuations in the short or medium term. Regrettably, numerous irrigation schemes in Cambodia lack hydraulic feasibility, proper design, and efficient operation. Climate change primarily impacts the hydraulic feasibility, design (of new or upgraded schemes), and operation, which rely on data and knowledge support.

1. The Ministry of Water Resources, River Development, and Ganga Rejuvenation (MoWR, RD & GR) has implemented short-term, medium-term, and long-term plans to address water resource development and management issues, with completion targets set for 2020, 2025, and 2035, respectively (Patel and Prajapati, 2020).
2. In the short term, actions are underway through the Pradhan Mantri Krishi Sinchayee Yojana (PMKSY), a comprehensive initiative aimed at facilitating the convergence of investments in irrigation. This initiative aims to combine District and State irrigation plans, providing comprehensive solutions throughout the entire irrigation

process, including water resource management, distribution, efficient usage, and extension services.

3. The medium-term strategy will prioritize addressing projects that have faced delays due to various factors and are currently not covered under the PMKSY.
4. The overarching goal involves the establishment of fresh initiatives, emphasizing the management of surface and groundwater resources, which encompasses projects like river interlinking. This initiative aims to enhance water utilization capacity from both surface and groundwater sources.

Case Study

Odisha Integrated Irrigation Project for Climate Resilient Agriculture (OIIPCRA)

The Odisha government, led by the Department of Water Resources (DoWR), has launched the "Odisha Integrated Irrigation Project for Climate Resilient Agriculture (OIIPCRA)." This initiative backed by the World Bank, centers on tank-based irrigation systems. It aims to demonstrate climate-smart agricultural techniques within an operational zone encompassing roughly 56,400 hectares across 538 minor irrigation projects. Furthermore, it aims to improve the capabilities and incomes of farmers in the project area by targeting around 70,000 hectares of rain-fed areas adjacent to these Minor Irrigation Projects.

Project Objectives

The project aims to provide technical assistance in several areas:

- Enhancing crop productivity to ensure food security and increase income.
    1. Supporting farmers in adopting resilient agricultural practices and technologies, for adaptation to and mitigation of climate change.
    2. Promoting diversification, particularly during the Rabi (winter) season and in certain areas cultivating upland rice during Kharif

(monsoon), to boost income, enhance nutrition, and adapt to climate change.

3. Implementing more efficient water-use practices and enhancing the quality and reliability of irrigation services.
4. Establishing a better framework for the efficient and sustainable management of surface and groundwater resources, including understanding sustainable levels of groundwater usage based on location-specific factors, providing systematic support to Participatory Planning (PPs), and piloting Integrated Water Resources Management (IWRM).
5. Improving produce marketing in 15 districts of Odisha.

OIIPCRA Implementing Authority/Executing Agency

The project is executed through cooperation among three government departments: Water Resources (DoWR), Agriculture and Farmers' Empowerment (A&FE), and Fisheries and Animal Resources Development (F&ARD). The "Odisha Community Tank Development & Management Society (OCTDMS)" oversees project operations.

OIIPCRA Project Area

The project covers 15 districts in Odisha, including Boudh, Gajapati, Ganjam, Balasore, Bargarh, Bhadrakh, Bolangir, Kandhamal, Keonjhar, Nawarangpur, Nuapada, Mayurbhanj, and Sonepur, Jajpur, Kalahandi. These districts are situated within four river basins: Rushikulya, Budhabalanga, Baitarani, and Tel-sub basin.

OIIPCRA Project Cost & Financing and Period for Implementation

The projected cost of the project amounts to approximately US $226.50 million (equivalent to around INR 1,683 crores). The project commenced in December 2019 and is slated for completion within six years, concluding by December 2025.

OIIPCRA Project Beneficiaries

The project is expected to directly benefit about 125,000 smallholder households in 15 participating districts. Around 72,800 households will gain from improved irrigation and climate-smart crop production within the irrigated command area, while approximately 52,200 households will benefit outside this area. Additionally, about 3,000 households will benefit from enhanced aquaculture production, 20,000 households with surplus produce will gain from improved marketing support, and approximately 500 Agri-Entrepreneurs, including women, will receive training and capacity-building support.

Future Recommendations

Research and Development

Innovative Technologies

Invest in research to develop and implement innovative technologies such as sensor-based irrigation systems, precision agriculture tools, and climate-resilient crop varieties. This will help optimize water use efficiency, reduce resource wastage, and enhance crop productivity in changing climatic conditions.

Adaptation Strategies

Focus on studying and implementing adaptation strategies that can mitigate the impacts of climate change on irrigation systems. This may include exploring drought-resistant crops, water-saving irrigation techniques, and soil moisture monitoring technologies.

Capacity Building

Training Programs

Develop and conduct training programs, workshops, and extension services to educate farmers, water resource managers, and policymakers on the importance of climate-resilient irrigation practices. These capacity-building initiatives should emphasize the benefits of sustainable water management and provide practical guidance on implementation.

Knowledge Sharing

Facilitate knowledge-sharing platforms, demonstration farms, and farmer field schools to encourage peer-to-peer learning and exchange of best practices in climate-resilient irrigation. This collaborative approach can help build local capacity and foster community engagement.

Policy Support

Incentive Mechanisms

Introduce incentive mechanisms such as subsidies, tax breaks, and financial support for farmers and agricultural enterprises adopting climate-resilient irrigation practices. These policy measures can be used to offset initial investment costs and encourage the widespread adoption of sustainable water management techniques.

Regulatory Frameworks

Develop and enforce regulatory frameworks that promote water conservation, efficient irrigation practices, and sustainable land use planning. Incorporating climate resilience considerations into national water policies and agricultural strategies is essential for long-term sustainability.

Integration of Technology

Smart Irrigation Systems

Promote the integration of smart irrigation systems that utilize real-time data, weather forecasts, and soil moisture sensors to optimize irrigation scheduling and water distribution. Leveraging technologies like artificial intelligence and the Internet of Things (IoT) can enhance precision irrigation and resource efficiency.

Remote Monitoring

Implement remote monitoring and control systems that enable farmers to remotely manage irrigation operations, monitor crop health, and detect water stress conditions. These technologies can improve decision-making processes and reduce water wastage in agricultural practices.

Collaborative Partnerships

Multi-Stakeholder Engagement

Foster partnerships between government agencies, research institutions, civil society organizations, and private sector entities to collaborate on climate-resilient irrigation projects. Pooling resources, expertise, and knowledge can accelerate the adoption of sustainable water management practices and scale up successful interventions.

Knowledge Exchange

Facilitate knowledge exchange platforms, research networks, and joint initiatives that promote cross-sectoral collaboration and learning. Encouraging dialogue and cooperation among diverse stakeholders can lead to innovative solutions and holistic approaches to climate-resilient irrigation.

Climate-Smart Agriculture

Integrated Approaches

Advocate for adopting climate-smart agriculture practices that combine climate-resilient irrigation with soil conservation, agroforestry, and biodiversity conservation measures. Integrating multiple strategies can enhance ecosystem resilience, improve agricultural productivity, and mitigate climate risks.

Adaptive Management

Encourage adaptive management approaches that allow farmers to adjust their irrigation practices in response to changing climatic conditions. Promoting flexibility, innovation, and learning-by-doing can help build resilience in agricultural systems and enhance adaptive capacity.

Monitoring and Evaluation

Performance Metrics

Develop clear indicators, benchmarks, and monitoring frameworks to assess the performance of climate-resilient irrigation interventions. Regular monitoring and evaluation activities can track progress, identify challenges, and inform evidence-based decision-making for continuous improvement.

Impact Assessment

Conduct rigorous impact assessments to evaluate the social, economic, and environmental outcomes of climate-resilient irrigation projects. Understanding the benefits, trade-offs, and unintended consequences of interventions is essential for refining strategies and maximizing positive impacts.

Community Engagement

Participatory Approaches

Engage local communities, farmers, and water users in the design, planning, and implementation of climate-resilient irrigation initiatives. Incorporating local knowledge, preferences, and needs into decision-making processes can enhance project relevance, ownership, and sustainability.

Empowerment

Empower farmers through capacity-building activities, farmer-led research, and participatory decision-making processes that enable them to take ownership of climate-resilient irrigation practices. Building trust, fostering collaboration, and promoting inclusive governance structures can strengthen community resilience and adaptive capacity.

By implementing these detailed future recommendations, stakeholders can collaborate to develop a more climate-resilient agriculture sector, enhance water resource management, and safeguard global food production in the face of climate change challenges.

Conclusion

In conclusion, the imperative of climate-resilient irrigation practices in agriculture cannot be overstated in the face of a changing climate. As global temperatures rise and weather patterns become increasingly unpredictable, the need to adapt irrigation techniques to ensure food security and environmental sustainability is more critical than ever. This chapter has underscored the importance of embracing innovative technologies, capacity-building initiatives, policy support, and collaborative partnerships to enhance water efficiency, mitigate climate risks, and promote resilient agricultural practices.

The exploration of various resilient irrigation methods, such as drip and sprinkler irrigation, alongside water management approaches like rainwater harvesting and recycling, highlights the diverse strategies available to improve water use efficiency and counteract the adverse effects of climate change on agriculture. Technological innovations, including remote sensing, GIS, weather forecasting, and smart irrigation controllers, offer promising avenues for optimizing irrigation processes and enhancing resource management in the face of evolving climatic conditions.

Effective policy and governance frameworks are central to fostering climate-resilient irrigation practices. By aligning government initiatives, water resource management strategies, legal frameworks, and international collaborations, stakeholders can create an enabling environment for the widespread adoption of sustainable water management practices. Real-world case studies highlighted successful implementations of climate-resilient irrigation, demonstrating the tangible benefits of integrating resilience-building measures into agricultural systems.

Looking ahead, the future of climate-resilient irrigation hinges on continued research and development, capacity-building efforts, policy support, technological integration, collaborative partnerships, climate-smart agriculture practices, monitoring and evaluation mechanisms, and community engagement. By embracing these recommendations and working collectively towards sustainable water management, stakeholders can navigate the challenges posed by climate change, safeguard water resources, and ensure the resilience of global food production systems in the years to come.

In essence, climate-resilient irrigation is not just a necessity but a pathway towards building a more sustainable and adaptive agricultural sector that can thrive in a changing world. By prioritizing resilience, innovation, and collaboration, we can pave the way for a more secure and sustainable future for agriculture and the communities that depend on it.

Disclaimer

None.

References

Adams, R. M., Brian H., Hurd., Stephanie Lenhart (1998) *Neil Leary Effects of global climate change onagriculture: an interpretative review* Vol. 11: 19–30, Climate Research.

Agrivi. (2023). Center pivot irrigation: from dust bowl to modern era methods. *AGRIVI.* https://www.agrivi.com/blog/center-pivot-system-an-efficient-and-economical-solution-for-irrigation/.

Bezas, K., & Filippidou, F. (2023). The Role of Artificial Intelligence and Machine Learning in Smart and Precision Agriculture. *Indonesian Journal of Computer Science*, 12(4), 1576-1588.

Bhuiyan, C., Flügel, W. A., and Singh, R. P. 2009. Erratic Monsoon, Growing Water Demand, and Declining Water Table. 2009. *Journal of Spatial Hydrology* Vol.9, No.1 Spring.

Chamara, N., Islam, M. D., Bai, G. (Frank), Shi, Y., & Ge, Y. (2022). Ag-IoT for crop and environment monitoring: Past, present, and future. *Agricultural Systems,* 203, 103497. https://doi.org/10.1016/j.agsy.2022.103497.

Chowdary, A. V., Saini N., Kumar A., Kumar S., Ballabh J., Bhatt S. S., Bhatt A., Prakash S, & Patel A. Influence of integrated nutrient management on physiological parameters of lentil (Lens culinaris Medik.). *Plant Science Today*, 2023;10(3), 94–97. https://doi.org/10.14719/pst.2124.

Devi, N., Sarma, K. K., & Laskar, S. (2023). Design of an intelligent bean cultivation approach using computer vision, IoT and spatio-temporal deep learning structures. *Ecological Informatics,* 75, 102044. https://doi.org/10.1016/j.ecoinf.102044.

Dimitriadis, S., & Goumopoulos, C. (2008). Applying machine learning to extract new knowledge in precision agriculture applications. In *2008 Panhellenic Conference on Informatics* (pp. 100–104). IEEE.

Feng, Y., Peng, Y., Cui, N., Gong, D., & Zhang, K. (2017). Modeling reference evapotranspiration using extreme learning machine and generalized regression neural network only with temperature data. *Computers and Electronics in Agriculture.*

Ghiassi, M., Fa'Al, F., & Abrishamchi, A. (2017). Large metropolitan water demand forecasting using DAN2, FTDNN, and KNN models: A case study of the city of Tehran, Iran. *Urban Water Journal,* 14(6), 655–659. https://doi.org/10.1080/1573062X.2016.1266152.

Goap, A., Sharma, D., Shukla, A. K., & Krishna, C. R. (2018). An IoT-based smart irrigation management system using machine learning and open-source technologies. *Computers and Electronics in Agriculture,* 155, 41–49.

Hirabayashi, Yukiko, Kanae, Shinjiro, Emori, Seita, Oki, Taikan and Kimoto, Masahide, 2008, "Global projections of changing risks of floods and droughts in a changing climate," *Hydrological Sciences Journal,* 53, 4, 754-772, doi: 10.1623/hysj.53.4.754.

International Commission on Irrigation & Drainage; 2018. Available:https://icidciid.org/publication/info/33.

IPCC. 2013. Climate Change 2013: The Physical Science Basis. Summary for Policymakers, Technical Summary and FrequentAsked Questions. In: T. F. Stocker, D. Qin, G. K. Plattner, M. M. B. Tignor, S. K. Allen, J. Boschung, A. Nauels, Y. Xia,

V. Bexand P. M. Midgley (Ed.), *Part of the Working Group I Contributionto the Fifth Assessment Report of the Intergovernmental Panelon Climate Change*, Geneva, Switzerland.

Islam, N., Rashid, M. M., Pasandideh, F., Ray, B., Moore, S., & Kadel, R. (2021). A Review of Applications and Communication Technologies for Internet of Things (IoT) and Unmanned Aerial Vehicle (UAV) Based Sustainable Smart Farming. *Sustainability,* 13(4), Art. no. 4. https://doi.org/10.3390/su13041821.

Jain, S. K. (n.d.). *Water Resource Management in India.* National Institute of Hydrology, Roorkee.

Kumar, A., Das N., Sharma H., & Singh R. Impact of Agrochemicals on the Environment and Ecological Alternatives. In A. Karmaoui (Ed.), *Water-Soil-Plant-Animal Nexus in the Era of Climate Change* 2024;317-328. IGI Global. https://doi.org/10.4018/978-1-6684-9838-5.ch015.

Kumar, A., Kumar S., Juyal R., Sharma H., Bisht M. Nanoscience in Agricultural Steadiness. In: Dar, G. H., Bhat, R. A., Mehmood, M. A. (eds) Microbiomes for the Management of Agricultural Sustainability. Springer, *Cham.* 2023;285-296. https://doi.org/10.1007/978-3-031-32967-8_17.

Kumar, A., Sharma, H., Shakeb, A., & Joshi, M. (2022). Environmental Sustainability: An Emerging Concept. *International Year of Millets* 2023, 29.

Kumar, R., S. Sachan, R. Kumar, K. Parmar, A. K. Mishra, A. Kumar and A. K. Mishra. 2023. Unveiling Use of Information and Communication Technology (ICT) in Western Uttar Pradesh, India: The Farmers' Perspective. *Journal of Global Innovations in Agricultural Sciences* 11:521-531. https://doi.org/10.22194/JGIAS/23.1197.

Kumar, R., Singh, A., Yadav, R. B., Kumar, A., Kumar, S., Shahi, U. P., & Singh, A. P. (2017). Growth, development and yield response of rice (Oryza sativa L.) as influenced by efficient nitrogen management under subtropical climatic condition. *Journal of Pharmacognosy and Phytochemistry,* 6(6S), 791-797.

Lester, R. and Gurenko, E., 2003, "Financing Rapid Onset Natural Disaster Losses in India: A Risk Management Approach," *World Bank, Report* No. 26844 IN, August 2003, 10-25.

Lu, Y. C., Sadler, E. J., & Camp, C. R. (2005). Economic Feasibility Study of Variable Irrigation of Corn Production in Southeast Coastal Plain. *Journal of Sustainable Agriculture*, 26, 69–81.

Meehl, G. A., Stocker, T. F., Collings, W. D., Friedingstein, P., Gaye, A. T., Gregory, J. M., Kitoh, A., Knutti, R., Murphy, J. M., Noda, A., Raper, S. C. B., Watterson, I. G., Weaver, A. J. and Zhao, Z. C., 2007, "*Contribution of working Group I to the fourth assessment report of the intergovernmental panel on climate change*," [Solomon, S., D. Qin, M. Manning, Z. Chen, M. Marquis, K. B. Averyt, M. Tignor and H. L. Miller (eds.)]. Cambridge University Press, Cambridge, United Kingdom and New York, NY, USA.

Mikhaylov, A., Moiseev, N., Aleshin, K. and Burkhardt, T. (2020). Global climate change and greenhouse effect. *Entrepreneurship and Sustainability,* 7(4), 2897.

Mohammadi, K., Shamshirband, S., Motamedi, S., Petkovic, D., Hashim, R., & Gocic, M. (2015). Extreme learning machine based prediction of daily dew point temperature.

Computers and Electronics in Agriculture, 117, 214-225. https://doi.org/10.1016/j.compag.2015.08.008.

Mohapatra, J. B., Jha, P., Jha, M. K., & Biswal, S. (2021). Efficacy of machine learning techniques in predicting groundwater fluctuations in agro-ecological zones of India. *The Science of the Total Environment,* 785, 147319. https://doi.org/10.1016/j.scitotenv.2021.147319.

Orlowsky, B. and Seneviratne, S. I., 2011, "Global changes in extreme events: Regional and seasonal dimension," *Climatic Change,* 110, 3-4, 669-696.

Patel, N. H., & Prajapati, C. R. (2020). Agricultural sprinkler for irrigation system. *International Journal of Engineering and Technical Research,* 9, 162-166.

Pathmudi, V. R., Khatri, N., Kumar, S., Abdul-Qawy, A. S. H., & Vyas, A. K. (2023). A systematic review of IoT technologies and their constituents for smart and sustainable agriculture applications. *Scientific African,* 19, e01577. https://doi.org/10.1016/j.sciaf.e01577.

Rajendran, K. and Kitoh, A., 2008, "Indian summer monsoon in future climate projection by a super high-resolution global model," *Current Science*, 95, 11, 1560-1569.

Sadras, V. O., Villalobos, F. J., & Fereres, E. (2016). Crop Development and Growth. In F. Villalobos & E. Fereres (Eds.), *Principles of Agronomy for Sustainable Agriculture.* Springer.

Sanchez-Lugo C., Morice, J., Nicolas, P and Arguez, A. (2019). Global surface temperature in 'State of the climate' Bull. Amer. *Meteor,* 101(8), S24–S26. https://doi.org/10.1175/ BAMS-D20-0104.

Senthil Vadivu, M., Purushotham Reddy, M., Rane, K., Kumar, N., Karthikayen, A., & Behare, N. (2023). An IoT-Based System for Managing and Monitoring Smart Irrigation through Mobile Integration. *Journal of Machine and Computing,* 3(3).

Sharma, H., Kumar A., Singh R, & Kumar S. Impact of Nanopriming and Omics-Based Applications for Sustainable Agriculture. In A. Singh, V. Rajput, K. Ghazaryan, S. Gupta, & T. Minkina (Eds.), *Nanopriming Approach to Sustainable Agriculture* 2023;175-191. IGI Global. https://doi.org/10.4018/978-1-6684-7232-3.ch008.

Shilpa, A., Muneeswaran, V., & Rathinam, D. K. (2019). A precise and autonomous irrigation system for agriculture: IoT based self-propelled center pivot irrigation system. In *2019 5th International Conference on Advanced Computing & Communication Systems (ICACCS)* (pp. 533-538). IEEE.

Shivanna, K. R. (2022). Climate change and its impact on biodiversity and human welfare. *Proceedings of the Indian National Science Academy,* 88(2), 160-171.

Singh, I. K. 2023. India's Agriculture Is Feeling the Effects of Extreme Weather Events. *The India Cable.* https://thewire.in/agriculture/indias-agriculture-is-feeling-the-effects-of-extreme-weather-events.

Singh, N., Mulye, S. S., and Pant, G. B. (1992) Some Features of the Arid Area Variationsover India: 1871-1984, *PAGEOPH,* 138(1), 135–150.

Singha, S., Pasupulet, S., Singha, S. S., Singh, S., & Kumar, S. (2021). Prediction of groundwater quality using efficient machine learning technique. *Chemosphere,* 276, 130265. https://doi.org/10.1016/j.chemosphere.2021.130265.

Sun, A. Y., Scanlon, B. R., Save, H., & Rateb, A. (2021). Reconstruction of GRACE Total Water Storage Through Automated Machine Learning. *Water Resources Research,* 57(2). https://doi.org/10.1029/2020WR028666.

Swaminathan, M. S. and Rengalakshmi, R. 2016. Impact of extreme weather events in Indian agriculture: Enhancing the coping capacity of farm families. *MAUSAM*, 67, 1; 1-4.

Vijai, C., Worakamol, W., & Elayaraja, M. (2023). Climate change and its impact on agriculture. *International Journal of Agricultural Sciences and Veterinary Medicine*, 11(4), 1-8. https://doi.org/10.25303/1104ijasvm0108.

Vohra, K., & Saxena, P. K. (2019). Irrigation growth in India – Prospects, initiatives and challenges. *Ministry of Water Resources, River Development and Ganga Rejuvenation, Government of India.*

Wassmann, R., G. C. Nelson, S. B. Peng, K. Sumfleth, S. V. K. Jagadish, Y. Hosen and M. W. Rosegrant. 2010. Rice andglobal climate change. In: S. Pandey, D. Byerlee, D. Dawe, A. Dobermann, S. Mohanty, S. Rozelle and B. Hardy (Ed.) *Ricein the Global Economy: Strategic Research and Policy Issuesfor Food Security*. Ch. 4.1 International Rice Research Institute, Manila, Philippines, pp. 411-432.

Yang, P., Wu, L., Cheng, M., Fan, J., Li, S., Wang, H., & Qian, L. (2023). Review on Drip Irrigation: Impact on Crop Yield, Quality, and Water Productivity in China. *Water*, 15, 1733. https://doi.org/10.3390/w15091733.

Yu, H., Wen, X., Wu, M., Sheng, D., Wu, J., & Zhao, Y. (2022). Data-based groundwater quality estimation and uncertainty analysis for irrigation agriculture. *Agricultural Water Management*, 262, 107423. https://doi.org/10.1016/j.agwat.2021.107423.

Zou, X., Li, Y. E., Gao, Q., & Wan, Y. (2012). How water saving irrigation contributes to climate change resilience - a case study of practices in China. *Mitigation and Adaptation Strategies for Global Change,* 17, 111-132. https://doi.org/10 1007/ s11027-011-9316-8.

Biographical Sketches

Jitendra Singh

Jitendra Singh is presently working as Assistant professor-cum-Junior scientist in the Department of Soil Science and Agricultural Chemistry in Veer Kunwar Singh College of Agriculture (Bihar Agriculture University, Sabour, Bhagalpur) Buxar Bihar. He has around 10 years of Teaching and Research experience. Previously he worked as Assistant Professor in various prestigious universities and also worked as Research associate in WTC, IARI, New Delhi. He has more than 20 Research publications in peer reviewed journals and authored one book and many book chapter. He has earned his M.Sc and Ph.D degree in Soil Science from Sardar Vallabhbhai Patel University of

Agriculture and Technology, Modipuram, Meerut. He has also qualified ICAR- NET. He has done his B.SC. Agriculture from RBS college, Bichpuri, Agra. He has also received Best Young Scientist Award and Best Teacher Award.

Atin Kumar

Atin Kumar is currently working as Assistant Professor at School of Agriculture, Uttaranchal University Dehradun, Uttarakhand, India. He holds a B.Sc. in Agriculture and M.Sc. in Agriculture with a specialization in Agricultural Chemistry & Soil Science from CCS University, Meerut. He obtained his Ph.D. in Soil Science from Sardar Vallabhbhai Patel University of Agriculture & Technology, Meerut (U.P.) in 2019. He has actively participated in numerous national and international workshops, conferences, and training. His expertise lies in soil fertility, nutrient management, and water quality. With almost 4 years of experience in teaching and research, he has published over 40 research papers, articles, abstracts, and book chapters. Atin Kumar honored with the Young Researcher Award, and the Global Scientist Award for his contributions to agriculture and sustainable development.

Nitish Karn

Nitish Karn is currently pursuing in M.Sc. Agriculture (Agronomy) at School of Agriculture, Uttaranchal University Dehradun, Uttarakhand, India. He has participated in Global millets (Shree anna) conference 2023 and national conference on Prospects and Challenges: Promoting Sustainable Farming System in Mountain Regions of Uttarakhand in 2024. His expertise lies in soil fertility, water management and nutrient management. He has published over various research papers, articles, abstracts, and book chapters.

Rohit Kumar

Rohit Kumar was born on 08 January 1992 at District Bulandshshr, U.P. He has obtained M.Sc. & Ph.D. (Agronomy) from, Sardar Vallabhbhai Patel University of Agriculture & Technology (SVPUA&T), Meerut (U.P.) in 2016 and 2019. He has also qualified ICAR/ASRB NET exam in 2017. Rohit Kumar served as assistant professor at Chandigarh University, Punjab, and SGT University, Gurugram, Haryana, respectively. He has published more than 15 research papers, articles, abstract, and book chapters. He has also received Young Professional Award-2018" and Young Scientist Award 2019. His area of expertise is precision water and nutrient management in field crops. He has nearly 4 years of experience in teaching, research, consultancy and agricultural extension. He is also involved in teaching undergraduate and postgraduate students and supervised about 4 M.Sc. Agriculture (Agronomy) students. He is currently working as Assistant Professor at Faculty of Agricultural Science, GLA, University, Mathura, Uttar Pradesh.

Amir Parnian

Amir Parnian is currently working at National Salinity Research Center (NSRC), Agricultural Research Education and Extension Organization (AREEO), Yazd, Iran. He has actively participated in numerous national and international workshops, conferences, and training. His expertise lies in soil salinity, nutrient management, soil salinization. He has published various research papers, articles, abstracts, and book chapters in Scopus and WoS indexed journals. Amir Parnian honored with various Award and actively serve as editor and reviewer in various journals.

Mina Mahbod

Mina Mahbod is currently working at Department of Educational Sciences, Farhangian University, Tehran, Iran. He has actively participated in numerous national and international workshops, conferences, and training. His expertise lies in soil salinity, nutrient management, soil salinization. He has published various research papers, articles, abstracts, and book chapters in Scopus and WoS indexed journals. Mina Mahbod honored with various Award and actively serve as editor and reviewer in various journals.

Mohamed A. E. AbdelRahman

Mohamed A. E. AbdelRahman is currently working as Associate Professor at Division of Environmental Studies and Land Use (DESLU), National Authority for Remote Sensing and Space Sciences (NARSS), Egypt. He holds a B.Sc. and M.Sc. degree in Agriculture Sciences from Faculty of Agriculture, Zagazig University, Egypt in year 2004 and 2009. He obtained his Ph.D. degree in Soil Science and Agricultural Chemistry from UAS University of Agricultural Science, GKVK, Bangalore, Karnataka, India in the year 2014. He has actively participated in numerous national and international workshops, conferences, and training. His expertise lies in Soil Survey, Land Degradation, GIS and Remote Sensing Technology and Soil Fertility. He has reviewer and member of editiorial board in various international journals. He has member of various journals and societies. He has published various research papers, articles, abstracts, and book chapters in Scopus and Web of Science indexed journals. Mohamed A. E. AbdelRahman awarded with the European Space Agency Prize for the Best Poster at 12th Conference of the African Association of Remote Sensing and the Environment and Scientific Publication Award from the National Authority for Remote Sensing and Space Sciences.

Chapter 2

Managing Irrigation Systems under Climate Change: Assessing Crop Water Demand and Developing Water-Saving Solutions

V. P. Kovalchuk[1,*], D.Sc.
H. A. Balykhina[2], PhD
P. I. Kovalchuk[2], D.Sc.
M. V. Yatsiuk[3], PhD
O. P. Voitovych[2], PhD
and D. O. Demchuk[1]
[1]Department of Information Technologies and Marketing Innovations,
Institute of Water Problems and Land Reclamation of NAAS, Kyiv, Ukraine
[2]Department of Ecology and Nature Use,
Institute of Water Problems and Land Reclamation of NAAS, Kyiv, Ukraine
[3]Director of Institute of Water Problems and Land Reclamation of NAAS, Kyiv, Ukraine

Abstract

Due to the increased frequency of droughts and temperature rises caused by climate change, the water demand for growing crops increases. For these conditions, system management in irrigation is proposed, which ensures effective decision-making on watering, rational use, and preservation of available water resources.

An original method of comparing and evaluating the growth of water demand of crops when increasing climate aridity is presented. The method is based on comparing water balance deficits ranked by the years

* Corresponding Author's Email: volokovalchuk@gmail.com

In: Irrigation Management
Editors: Maha Elbana and Youssef Brouziyne
ISBN: 979-8-89530-318-4

of estimated water availability in two-time intervals over the last decades (1960-1990 and 1991-2021). To implement the method of assessing the increase in climate aridity, curves of water availability were constructed by the deficit of the water balance for alfalfa for the above periods using the data of the weather stations of Ukraine - MS Odesa, MS Izyum (Kharkiv region).

Using instrumental and computational methods, a structural and functional scheme for the field condition monitoring model and irrigation decision-making was developed; the new instrumental and computational approach combines irrigation management based on soil moisture measurement with forecasting and irrigation calculation based on weather service data. It is planned to create a database of water and physical soil properties and parameters of biologically optimal and water-saving irrigation regimes.

An instrumental and computational algorithm for making irrigation decisions was developed to implement an irrigation management system. Instrumental methods using an automated tensiometer and Watermark provide feedback on daily soil moisture status in the field. The method of identification of sensor readings of soil water potential depending on the type and temperature of the soil is theoretically and experimentally grounded. Irrigation forecasting is carried out using Internet weather services by the computational method. Optimal irrigation control occurs in a “sliding” time interval. A formalized algorithm for determining irrigation terms and rates based on predictive calculations is given.

By combining the algorithm of the instrumental and calculation approach with biologically optimal irrigation regimes, effective use of water resources, accuracy, and efficiency of irrigation application is achieved. When using the parameters of water-saving irrigation regimes, water resources can be saved by up to 30%, which is relevant in climate change conditions.

Keywords: algorithm, instrumental-calculation method, irrigation control, irrigation rates, the water balance deficit, water-saving regimes

List of Abbreviations

API	Advanced Programming Interface;
BBCH	BBCH-scale from “Biologische Bundesanstalt, Bundessortenamt und CHemische Industrie” is proposed to identify the phenological development stages of plants;
FAO	Food and Agriculture Organization of the United Nations;

GWT	Ground Water Table represents the top of the water-saturated zone in the soil;
MMHC	Minimum Moisture-Holding Capacity is the water remaining in the soil after it has been thoroughly saturated and allowed to drain freely, usually for a few days (eq. Field Capacity);
MS	Meteorological Station from the State Hydrometeorological Network of Ukraine;
SPEI	Standardized Precipitation Evapotranspiration Index
SPI	Standardized Precipitation Index;
UAH	national currency of Ukraine.

Introduction

In conditions of climate change, particularly when climate aridity increases, it is necessary to adjust approaches to irrigation management. Management solutions that ensure more efficient use of available water resources and measures to cover water deficits are important.

For this purpose, our study proposed a method for estimating the change in water demand based on calculating the water balance deficit when irrigating the main crops. The water balance deficit was compared over the years of different estimated water supply in two-time intervals of about 30 years: 1960-1990 and 1991-2021. The first-time interval corresponds to the period when irrigation systems were designed and built in Ukraine and the first decades of their operation. The second interval covers the years, including the last decades, when an increase in temperature and a decrease in precipitation were observed. That led to a rise in the water deficit for crops, that is, an increase in the need for irrigation.

Irrigation system management is a multi-component and complex process of obtaining data and making decisions. In world practice, decision-making on irrigation is widely used based on the instrumental method of measuring soil moisture content with various sensors. Both in Ukraine and abroad, calculation methods for irrigation are also implemented during the development of irrigation management systems. One of the components of the decision-making process is the choice of the optimal irrigation regime, which can ensure the planned yield and water saving in climate change conditions.

So, the problem is to estimate the crop water demand to develop instrumental and calculation models and irrigation control algorithms, which, together with efficient irrigation regimes and proper structure of irrigated areas, can ensure resource and water savings.

Literature Review

Today, due to climate change both in Ukraine and in the whole world, an increase in temperature has become an indisputable fact that affects crop productivity, which is confirmed in many scientific works (Adamenko, 2019; Kovalchuk P. et al., 2021; Trnka et al., 2011; Wilson et al., 2021). An increase in temperature leads to an increase in evapotranspiration of crops, and accordingly, water needs change for crops. Determining crop water demand and comparing it with the available local water resources is shown in the literature (McKee, Doesken, and Kleist 1993; Palmer 1965; Sielianinov 1966; Vicente-Serrano, Beguería, and López-Moreno 2010). The ratio of precipitation to the potential evaporation is expressed by the hydrothermal coefficient (Sielianinov 1966) and Palmer's aridity index (Palmer 1965). The Standardized Precipitation Index (SPI) was developed in (McKee, Doesken, and Kleist, 1993) to quantify precipitation deficits using monthly precipitation data at time intervals of 1 to 48 months (the most commonly applied monthly intervals were 1, 3, 6, 12, 24). Based on SPI, the Standardized Precipitation Evapotranspiration Index (SPEI) was proposed by (Vicente-Serrano, Beguería, and López-Moreno, 2010), which expresses the water deficit, being a difference between precipitation and potential evapotranspiration in a given area. Applying the SPEI index (Yang et al., 2016) analyzed water scarcity in the Haihe River Basin. However, all these coefficients do not express a particular crop's water demand during its growing season in a specific area. Therefore, the authors of this study assessed the increase in water demand for some crops due to climate change in the Dnipro-Donbas Canal Zone (Ukraine) and gave it in (Kovalchuk P. et al., 2021). Thus, the increased crop water demand is compensated due to irrigation (Kovalchuk P. et al., 2019).

As said in the statement of the problem, various sensors (Bianchi et al., 2017; Kovalchuk V. et al., 2021; Lozoya et al., 2016; Payero et al., 2017; Thakur et al., 2019; Thalheimer, 2013) and their networks (Kovalchuk V. et al., 2021; Thakur et al., 2019), in particular, an automated tensiometer (Bianchi et al., 2017; Kovalchuk V. et al., 2021; Thalheimer, 2013), are used in global practice to monitor soil moisture in the field. Sensor readings of modern irrigation management systems are used to make irrigation decisions (Lozoya et al., 2016; Aruna et al., 2016; Gadzalo et al., 2019; Goap et al.,

2018; Liang et al., 2020). In the developed "Irrigation Online" system (Gadzalo et al., 2019), wireless monitoring of soil moisture and meteorological parameters (Kovalchuk V. et al., 2021) in the field is used as feedback in the control system for daily correction of soil moisture. Calculation balance models are also used to determine soil moisture when calculating evapotranspiration by various factors (Bouraima, Zhang, and Wei, 2015; Smith, Kivumbi, and Heng, 2000; Solangi et al., 2022; Zhovtonog et al., 2015). Therefore, combining the advantages of instrumental methods (constant feedback from the field) with the calculation ones is extremely urgent in modern conditions. Calculation methods provide irrigation forecasting (Kovalchuk V. et al., 2019) on a "sliding interval" (Toloknianenko, 1975; Ushakov and Vojtov, 2004) using Internet weather forecasting services, for example OpenWeatherMap or AccuWeather.

It is known that the cumulative indicator of the irrigation regime is the season irrigation rate. It expresses the total water amount applied to the field during the growing season. A comparative assessment of irrigation rates of main crops based on experimental studies (Kovalchuk P. et al., 2019) shows that water-saving regimes compared to biologically optimal regimes allow for saving water and energy resources in different years with minor crop losses.

The research aims to assess methods for estimating the rise in crop water demand caused by climate aridity. It seeks to integrate instrumental and calculation models and irrigation control methods to ensure resource conservation and water-saving irrigation regimes.

Methods and Models

Method for Assessing Crop Water Demand

A new method and models are proposed based on calculating crop water demand rates under irrigation, which allows for estimating their changes over several years under climate change in a particular territory. The calculation method is based on the annual curves of moisture availability by the deficit of crop water balance. The method compares these curves and assesses the total increase in crop water demand in the years of different water supply taken for a certain 30-year period (Kovalchuk P. et al., 2021).

Different methods of calculating total evapotranspiration are known. Shtoyko's biophysical method is the most adapted to the conditions of Ukraine

(Shtoiko et al., 1977; Kovalchuk P. et al., 2021; Kovalchuk V. et al., 2019). It was approved in Ukraine for operational irrigation planning (Methodical Guidelines for the Operational Planning of Irrigation Regimes, 2004). By Shtoiko's formulas, evapotranspiration is calculated, and the deficit of crop water balance is determined at key weather stations for the territory of Ukraine:

$$D_i = p_i - E_i,\ i = 1,...,l, \tag{1}$$

where D_i is water balance deficit, p_i is precipitation, and E_i is evapotranspiration.

The deficit of the water balance for crop growing season for a particular year j is calculated based on the general deficit of the water balance for the growing season, taking into account the specifics of modern technologies for crop growing:

$$\omega_j = \sum_{i=1}^{l} D_i,\ j = 1,...,N, \tag{2}$$

where l is the number of ten-day periods in the crop growing season, for example, the water balance deficit was calculated for alfalfa crops during the growing season from April 1 to October 10 from 1960 to 2021 (Figure 1a).

Considering a series of N years with the necessary total deficit ω and biologically optimal irrigation rates, we can observe their stochastic nature of change over many years. To obtain a series with a gradually increasing deficit, it is ranked by N years. The water supply of the year $P(\omega_j)$ is calculated by the total water balance deficit using the well-known formula of S.M. Alpatyev (Kovalchuk P. et al., 2019):

$$P(\omega_j) = \frac{j - 0{,}3}{N + 0{,}4} * 100\%, \tag{3}$$

where j is the year number in the ranked series; N is the number of years in a row.

A graph is made by the water balance deficit (Figure 1b), where the curve increases from the wettest to the driest year.

When studying the changes in the crop irrigation water demand due to climate change, the entire series of observations is divided into two periods:

N_1 is the number of years of the first period, and N_2 is the number of years of the second period.

The ranking of the number of years in each period is made by the water balance deficit in the order of their growth for a particular crop:

$$\omega_1 \leq \omega_2 \leq \ldots \leq \omega_{N_1} \tag{4}$$

$$\omega_{N_1} \leq \omega_{N_1+1} \leq \ldots \leq \omega_{N_2} \tag{5}$$

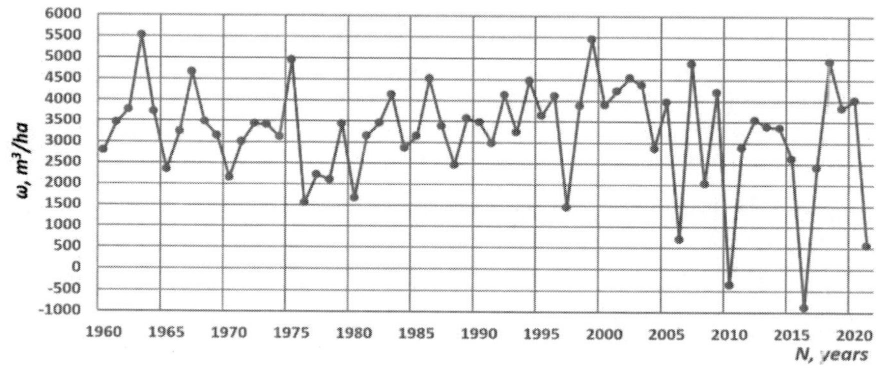

(a)

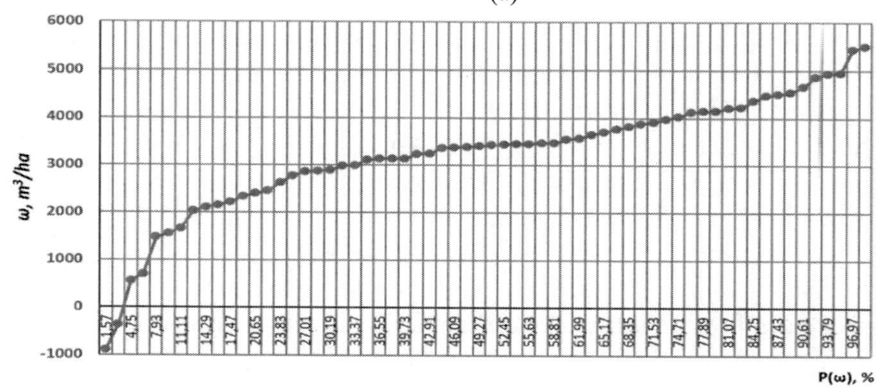

(b)

Figure 1. Water balance deficit for alfalfa, calculated by the Odessa weather station for 1960-2021: a: in ascending order of years; b: when ranking in order of increasing deficit.

The curves of water supply over the year are built by the water balance deficit for each period (N_1, N_2) of the study.

The trend of irrigation rate growth for individual years, for a group of years, or different periods is estimated by the difference in the areas S_1 and S_2, limited by the curves of the integral water supply of the year for the water balance deficit:

$$M(\Delta) = S_1 - S_2 = \int_1^{N_1} \omega_j dp(\omega_j) - \int_1^{N_2} \omega_j dp(\omega_j). \tag{6}$$

The difference $M(\Delta)$ expresses the long-term average change in irrigation rate due to climate change or the corresponding mathematical expectation of this difference.

Operational Irrigation Control Model

The model provides a structural and functional diagram of obtaining data and making irrigation decisions (Figure 2). The database of water-physical properties of the soils of Ukraine was created based on the Van Genuchten model (Van Genuchten, 1980), the USDA ROSETTA Model and the methodical approach (Kovalchuk V., 2021). A knowledge base on irrigation rates, antecedent soil water thresholds, and other parameters of various irrigation regimes (biologically optimal, water-saving, soil-protective, etc.) under climate change was also created (Kovalchuk P. et al., 2019). The field conditions were automatically monitored for operational irrigation calculations, as described in the work (Kovalchuk P. et al., 2021).

A combination of instrumental input parameters and calculation data for decision-making is assumed. Irrigation decisions are made by applying a special algorithm that uses field instrumental measurements and models for calculating the balance and predicting irrigation on a "sliding" interval using the data from meteorological Internet services.

Irrigation Regimes Given Water-Saving

The problem of adaptation and selection of parameters for irrigation regimes is solved based on the application of certain criteria and requirements for irrigation control in specific conditions of water saving. The type of irrigation regime is chosen at the beginning of the irrigation season when drawing up business plans for farms depending on the irrigation strategy. When there is no water deficit, there are optimal conditions that fully meet the requirements of crops for soil water supply. Such conditions provide the best crop productivity. However, obtaining such productivity is not always covered by irrigation water costs. With a shortage of water resources, it is possible to plan optimal irrigation regimes for the areas where the crops of higher economic importance are cultivated.

In the conditions of water conservation, in case the farm adopts a strategy of saving water and logistical resources with a slight decrease in productivity, water-saving irrigation regimes are used (Methodical Guidelines for the Operational Planning of Irrigation Regimes, 2004).

Water-saving irrigation regimes are developed to meet the economic requirements for optimizing the additional net profit from irrigation (Kovalchuk P. et al., 2019). A necessary tool for water saving is the regulated use of water. The regulatory use of water in climate change conditions and the need for water conservation require planning irrigation when using water-saving irrigation regimes and determining the appropriate biologically optimal and economically justified irrigation rates. The biologically optimal irrigation regimes can also be applied when there is a sufficient amount of water.

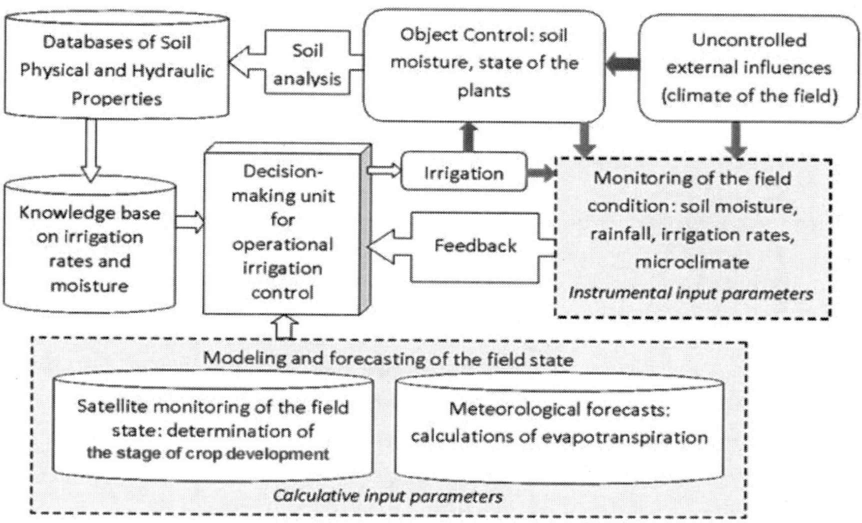

Figure 2. Structural and functional diagram of the model for making decisions on irrigation using the instrumental and calculation method.

When concluding contracts with agricultural producers for water supply, water suppliers must apply a preferential tariff for water supply when applying water-saving irrigation regimes, a regular tariff for biologically optimal irrigation regimes, and an increased tariff for overtime water supply.

At the same time, water consumers must be guided by economically justified design rates u_p. The determination of the economically justified rate is based on the economic evaluation of the variants of the design rate u_p by the criterion of additional net profit from irrigation (Kovalchuk P. et al., 2019):

$$F_3(P) = (C - C_1) \cdot f\left(\frac{u_p + p_j}{\omega_j + p_j}\right) \cdot Y^p - (C - C_2) \cdot f\left(\frac{p_j}{\omega_j + p_j}\right) \cdot Y^p - \lambda u_p, \quad (7)$$

where $F_3(P)$ is additional net profit from irrigation, UAH ha^{-1}; P is the design level of ensuring the estimated irrigation rate, % (determined by the formula (3)); C is the purchase price, UAH centner^{-1}; C_1, C_2 are cost prices, respectively, under irrigation and on dry land (without costs for water supply), UAH centner^{-1}; Y^p is planned (design) productivity, centner ha^{-1}; $f\left(\frac{u_p + p_j}{\omega_j + p_j}\right)$, $f\left(\frac{p_j}{\omega_j + p_j}\right)$ is productivity under irrigation and on dry land, in fractions of a unit; ω_j is the value of the current biologically optimal irrigation rate, m^3 ha^{-1}; u_p is designed water-saving irrigation rate, m^3 ha^{-1}; p_j is precipitation, m^3 ha^{-1}; λ is tariff (the price of 1 m^3 of water), UAH m^{-3}. The dependence of crop productivity on the water factor is given in Kovalchuk P. et al., (2019).

Results

Comparison of Changes in Crop Water Demand under Climate Change

The system irrigation control was studied to determine crop water demand in water shortage conditions. The graph (Figure 3) shows the curves of the water supply of the year due to the deficit of the water balance for each period (N_1, N_2) of the study when using formulas (3)-(5). The water balance deficit (Figure 3) when growing alfalfa for the 2nd and subsequent years was analyzed. Alfalfa has a pretty long growing season from April to early October, so it can be argued that the entire growing season should be considered (covered).

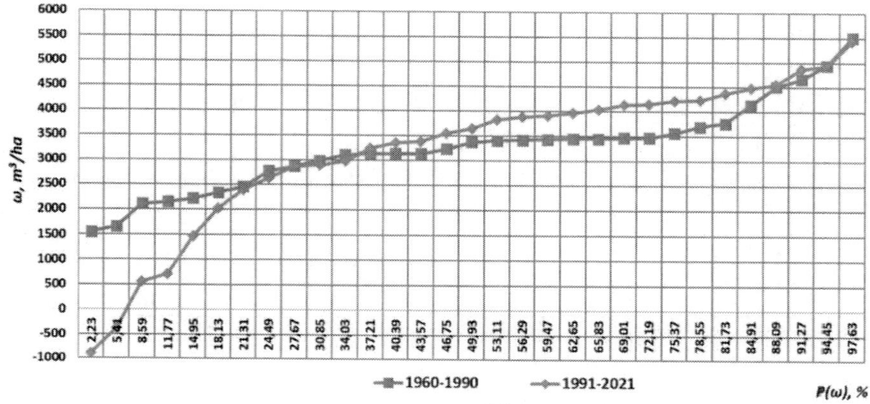

(a)

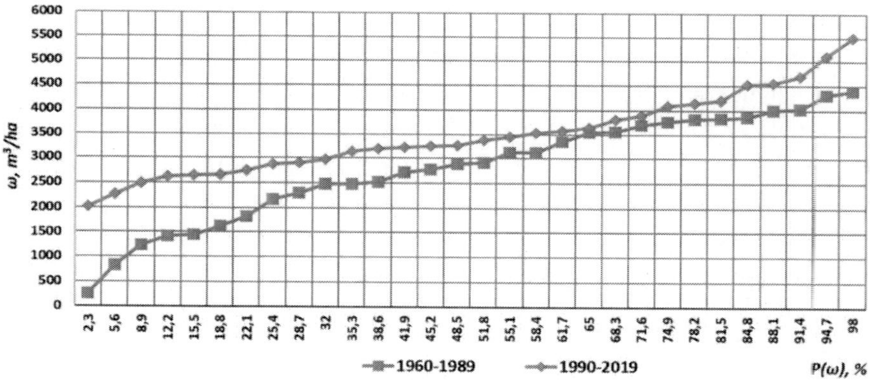

(b)

Figure 3. Curves of water supply by the water balance deficit (as in the case of alfalfa) for different periods: (a) by the data of the Odessa weather station and (ɔ) by the data of the Izyum weather station (Kharkiv region).

According to the Odesa Meteorological Station (MS) data (Figure 3a), the increase in water balance deficit is about 10-15% in dry and medium dry years. In wet and medium-wet years, the water balance deficit did not increase but decreased, which is explained by heavier precipitation in the Odesa region in recent years. However, because the average, medium dry, and extremely dry years, up to 90% probability, have become drier, it is necessary to apply higher irrigation rates to compensate for the water balance deficit.

Table 1. Comparison of climate change in different years by precipitation

Years by the amount of precipitation	*P, %*	The result of climate change	
		Odesa MS Southern Ukraine	Izyum MS Eastern Ukraine
extremely wet	0-20	became much wetter	became much drier
medium wet	20-40	almost did not change	became drier
average	40-60	became a little drier	became a little drier
medium dry	60-80	became much drier	almost did not change; changes are visible starting from >75%
extremely dry	80-100	became drier, but extremely dry years (>90%) did not almost change	became much drier

According to the data of Izyum MS (Figure 3b), a significant increase in water balance deficit was observed in extreme years. The dry years became even drier, and there were almost no wet years. So, the water balance deficit, which has to be compensated by irrigation for sustainable agricultural production, is now permanent.

The analysis of annual water supply curves in different years by the water balance deficit is given in Table 1.

Under climate change conditions, it is recommended that water-saving irrigation regimes be applied. These irrigation regimes are based on instrumental and calculation algorithms of irrigation control.

Algorithm of Operational Irrigation Control Based on the Instrumental and Calculation Method

Based on the research results, a combined irrigation control algorithm (Figure 4) on the “sliding interval” was developed. The algorithm combines instrumental methods of scheduling irrigation by soil moisture content with calculation methods of predictive soil moisture on the time interval τ when using Internet weather services. Sensor stations, based on telecommunication technology, are used for feedback (Figure 4).

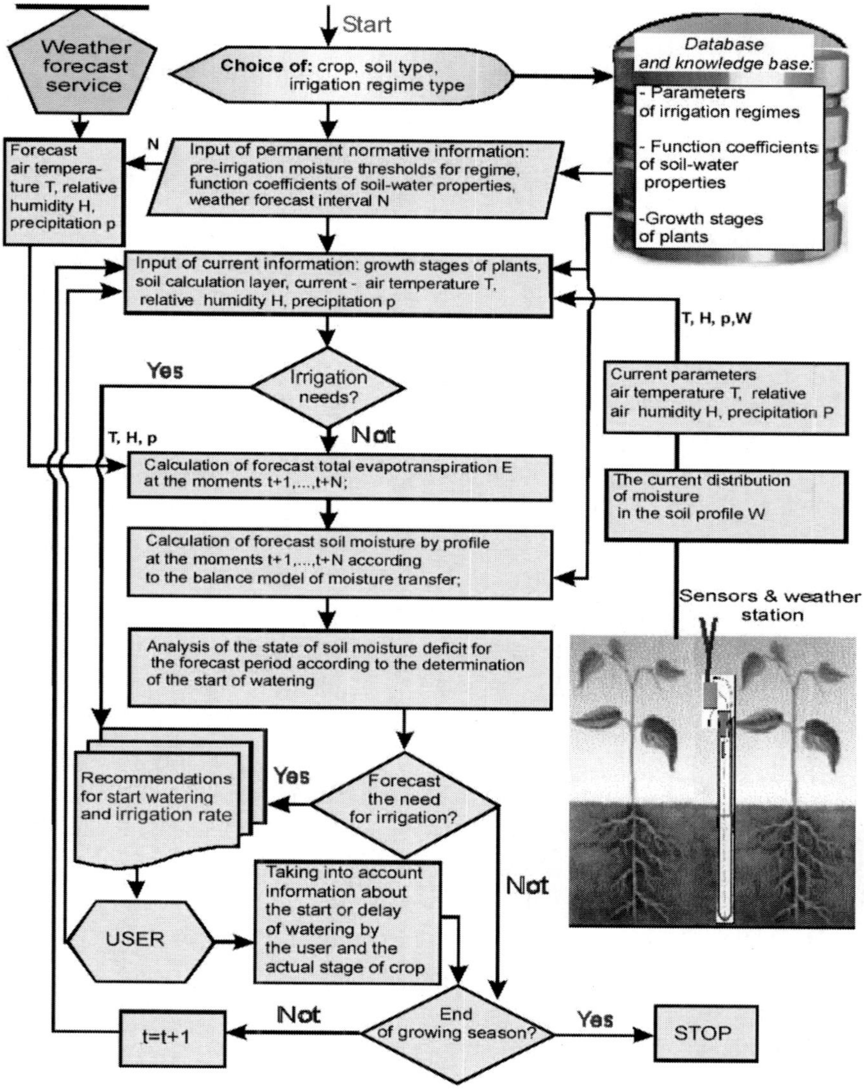

Figure 4. Block diagram of the algorithm of intelligent combined irrigation control on a "sliding" interval and various types of feedback in the control system.

After selecting a crop, soil type, and irrigation regime, constant normative information about the parameters of the selected irrigation regime flows from the database (the principles of obtaining irrigation rates are given in "Method for Assessing Crop Water Demand" and "Comparison of Changes in Crop Water Demand under Climate Change"), the coefficients of water-physical properties functions of the selected soil (they are described in "Operational Irrigation Control Model") and information about the stage of crop development. The next step in the algorithm execution is the flow of current information, including from field monitoring stations. At this step of the algorithm execution, the need for irrigation is determined by the instrumental method. The instrumental part of the control system consists of field weather stations and moisture meters (automatic tensiometers or Watermark sensors), which automatically receive data on soil moisture and atmospheric parameters at a certain point in the field and transmit them using telecommunication technology to the Internet server for further processing and calculations (Figure 4).

It must be noted that automatic tensiometers or Watermark sensors measure soil moisture capacity, and then these data are converted to volume moisture content using the Van Genuchten model (Van Genuchten, 1980) and the USDA ROSETTA Model. Irrigation at this algorithm step is scheduled based on data on volume moisture content, deposits of moisture, or moisture deficit in the estimated soil layer.

If, during the execution of the algorithm, a pre-irrigation threshold of moisture is not reached, then irrigation time is specified by the calculation method at the interval τ using the balance calculations. Based on the forecast data of the Internet Weather Forecast Service on the temperature, relative humidity, and precipitation, which are obtained automatically through API (Advanced Programming Interface), predictive evapotranspiration is calculated using the Shtoiko method (Shtoiko et al., 1977; Kovalchuk V. et al., 2019). The data from these calculations are used in forecasting soil moisture (Figure 4) using the balance method. The need for predictive irrigation calculations for making organizational decisions is due to the necessity of preparing irrigation equipment, purchasing water, etc.

A formalized process of irrigation control on a "sliding" interval (Toloknianenko, 1975; Ushakov and Vojtov, 2004) described by predictive calculations when using the balance method is given below. At some step n, the actual soil moisture or soil moisture deficit D_n is determined by the instrumental method. Then, to control irrigation on a "sliding" interval $t \in \tau = [n; n+N], N \geq 2,$ the predictive deficit of moisture capacity in the soil is determined by the balance model:

$$D_{t+1} = D_t - U_t + (\alpha E_t - p_t), t \in \tau = [n; n+N], N \geq 2, \tag{8}$$

where D_{t+1}, D_t – predictive moisture deficit in the soil at the beginning and the end of the t step of the calculation (it is one day in our study); E_t is evapotranspiration predicted for the time interval τ, is calculated by the data obtained from the meteorological Internet services; α is the coefficient of reduction of evapotranspiration due to suboptimal soil irrigation (Methodical Guidelines for the Operational Planning of Irrigation Regimes, 2004), p_t is the predictive amount of precipitation, U_t is the irrigation rate as a control parameter.

For the synthesis of the control system by equation (8), it is necessary to determine the values of the control parameters – the terms and rates of irrigation. When there is no precipitation, the deficit D_t constantly increases so that at a particular time $t=t^*$ on the interval $[n; n+N]^*$ soil moisture deficit becomes higher than the critical one, i.e., the constraint is executed

$$D_{t^*} \geq D_{cr}. \tag{9}$$

The control value U_{t^*} on a "sliding" interval is determined by:

$$U_{t^*} = m_{t^*}, \tag{10}$$

where m_{t^*} is the discrete irrigation rate determined for each stage of plant development; D_{cr} is the critical value of the deficit that requires irrigation.

The development stage is identified and predicted using the Internet weather information services or remote sensing (Kovalchuk et al., 2023). In the critical stages of crop development, a decrease in soil moisture below optimal values is dangerous as it can lead to significant crop losses.

Since with each new cycle n, the predictive interval is shifted, scheduling the irrigation with the rate m_{t^*} is performed taking into account new information about the object and predictive uncontrolled weather effects E_t and p_t, affecting it. Only the first unit of the control sequence U_{n+1} is introduced on the object, so the forecasting process allows tracking the date of the irrigation scheduling. A feature of the algorithm that ensures its effectiveness is applying the precipitation forecast of precipitation, temperature, and relative humidity.

Thus, the predictive need for irrigation is determined, and according to the algorithm, the user receives recommendations on the timing and rates of irrigation. So, the instrumental-calculation method allows not only scheduling irrigation by the instrumental method or current calculations but also preparing for irrigation by its forecast using the calculation method.

Introduction of Water-saving Irrigation Regimes

Kovalchuk P. et al., (2019) recommended using models and parameters of water-saving irrigation regimes in control systems at different levels of resource conservation.

The introduction of water-saving irrigation modes (water-saving regime I) is based on reducing the estimated soil layer and feeding it with groundwater. In balance calculations, in this case, evapotranspiration is determined by the formula:

$$E_p = K_r \alpha K E, \tag{11}$$

where E_p is the calculated evapotranspiration in the $h<1$ m layer; E is evapotranspiration under optimal conditions in the $h=1$ m layer; K_r is the coefficient of feeding of the calculated layer with groundwater; K is the coefficient of evapotranspiration reduction, if the calculated layer h is less than 1 m, and the root layer h_k is larger than the calculated one h $(h_k > h)$. Coefficient K is determined by the formula (Methodical Guidelines for the Operational Planning of Irrigation Regimes, 2004):

$$K = \begin{cases} \dfrac{h}{h_k}, & if\ h_k > h,\ h_k \le 1m; \\ 1, & if\ h_k < h,\ h \le 1m. \end{cases} \tag{12}$$

If the entire root system in a particular phase of plant development is in the calculated layer, then the total evaporation does not decrease, i.e., $K=1$.

In the case of close fresh and weakly mineralized groundwater, a part of the consumed moisture comes to the calculated layer from groundwater. In balance calculations, the amount of evapotranspiration should be multiplied by the Kr coefficient. Numerical values of reduction coefficients are given in Methodical Guidelines for the Operational Planning of Irrigation Regimes (2004).

If the calculated soil layer is reduced, then up to 15% of water is saved without actually reducing the yield (water-saving regime I). When reducing the pre-irrigation thresholds in non-critical phases of plant development, along with reducing the calculated soil layer (water-saving regime II), then water is saved up to 25-35%, and the yield decreases (up to 3-7%) (Table 2).

Practical Implementation of Irrigation Control Based on the Instrumental and Calculation Algorithm during the Growing Season

An example of irrigation control, when applying the biologically optimal irrigation regime by the algorithm described above, during the growing season was practically implemented in the Odessa region in 2022 (Figure 5).

Table 2. Irrigation rates of crops (net, in the field) and the number of irrigation when applying biologically optimal and water-saving irrigation regimes in years of different water supply, m^3 ha^{-1}

Crops	Dry and extremely dry years*			Average years		
	Biologically optimal	Water-saving		Biologically optimal	Water-saving	
		GWT > 3 m	1,5 < GWT < 3 m		GWT > 3 m	1,5 < GWT < 3 m
Winter wheat	2000-2400	800-1200	600-900	1200-1800	400-800	300-500
	5-6**	2-3	2-3	3-4	1-2	1-2
Corn for grain	3000-4500	1600-2300	1500-2100	1800-2900	1000-1700	1000-1200
	6-11	4-6	4-6	4-6	3-4	3-4
Soybean	3200-4200	1800-2400	1600-2200	2200-2800	1400-1800	1200-1500
	6-11	4-6	4-6	5-7	3-5	3-4
Alfalfa of the second year	4200-4700	2300-2600	1100-1500	2600-4000	1700-2100	800-1000
	8-11	6-7	4-5	5-8	5-8	3-4
Fodder beet	3600-4200	1900-2400	1200-1700	2400-3000	1600-2000	900-1100
	7-11	5-6	4-5	5-6	4-5	3-4

Note: GWT ground water table; * years of different water supply; **numbers of irrigations.

Studied crop - corn for grain, was sown on April 30 in heavy loam black earth soil. The water-physical properties of the soil were determined based on soil samples taken from the field using the methodical approach given in Kovalchuk (2021) using Van Genuchten's model (Van Genuchten, 1980) and the USDA ROSETTA Model.

The minimum moisture-holding capacity (MMHC) of the soil was determined by the methodology (Barros et al., 2017) and was - 214 mm or 2140 m^3 ha^{-1} (line 1, Figure 5a). Irrigation control is performed when applying the biologically optimal irrigation regime. Pre-irrigation thresholds of soil moisture are differentiated by the stages of corn development (line 2, Figure 5a). From the sowing stage to the stage of the 11th leaf, the moisture content was maintained at no lower than 75% of MMHC. Starting from the stage of the 11th leaf to the stage of milky ripeness, during the critical period of corn growth, the soil moisture was maintained at no lower than 80% of MMHC. In the stage of milky ripeness, soil moisture was maintained at no lower than 70% of MMHC; in the waxy ripeness stage, soil moisture was maintained at no lower than 65% of MMHC (line 3, Figure 5a).

Weather conditions of the growing season of 2022 due to the deficit of the water balance (as shown in Figure 3) can be attributed to extremely dry years (only 23.9 mm of precipitation fell from April to September). Therefore, when introducing the biologically optimal irrigation regime in the meteorological conditions of 2022, a seasonal irrigation rate was 4160 m^3 ha^{-1}. Nine irrigations with the rate of 400-500 m^3 ha^{-1} and one refreshing irrigation with the rate of 100 m^3 ha^{-1} were recommended in the phase of corn flowering (BBCH 60–69: Flowering, Anthesis), as the relative air humidity at that time was very low (39-47%).

Discussion of Results

In the conditions of increasing climate aridity, comparing crop water demand at different time intervals is recommended for assessing the change in crop demand for irrigation water (Kovalchuk P. et al., 2021). The water balance deficit was compared in years of the same water supply. The increase in crop water demand in Ukraine, given in "Comparison of Changes in Crop Water Demand under Climate Change" and in Kovalchuk P. et al., (2021), proved the need to develop a systematic approach to irrigation control. The systematic approach includes detailed monitoring of the state of the field and an algorithm for operative irrigation control based on instrumental and calculation methods.

Biologically optimal or water-saving irrigation regimes are used depending on the available water resources.

Precipitation has a stochastic character in semi-arid regions, including the south of Ukraine (Kovalchuk V. et al., 2019). Therefore, forecasts of precipitation are essential for the forecast of irrigation. In this case, the systematic approach compared to the study results when using short-term weather forecasting (Danlu et al., 2023; Perera et al., 2014; Perera et al., 2016), quantitatively reduces the risks of over- and under-irrigation and ensures water savings during irrigation. Irrigation is scheduled at a "sliding" interval using an instrumental method based on measuring the initial state of moisture in the soil. As a result, the risk of insufficient soil moisture content is reduced. Daily evapotranspiration for 5-7 days in advance and irrigation timing are determined by the calculation method according to the predictive precipitation, temperature, and air humidity (Kovalchuk V. et al., 2019; Perera et al., 2016).

This enables a farmer to prepare for irrigation and use precipitation efficiently, as suggested in the works (Danlu et al., 2023; Cao et al., 2023). The proposed algorithm is combined, as it combines instrumental determination of soil moisture and calculation methods of irrigation forecasting (Bouraima, Zhang, and Wei, 2015; Smith, Kivumbi, and Heng, 2000; Solangi et al., 2022). It is intelligent because it contains an online system of data transfer, forecasting, and decision-making under uncertainty.

Many researchers (Madhusudhan et al., 2021; Nath et al., 2022; Ewaid, Salwan, and Nadhir, 2019; Solangi et al., 2022) use the FAO Cropwat8 model for irrigation control. The Penman-Monteith evapotranspiration model (Allen et al., 1998) used in Cropwat8 has high requirements for input data that are not provided by Internet weather forecasting services. And as studies (Faybishenko, 2012; Pandey, Dabral, and Vanita Pandey, 2016; Yang et al., 2021) show, the accuracy of all methods of determining evapotranspiration is affected by the stochastic nature of climatic factors. Therefore, the main advantage of the proposed system approach is the presence of feedback in the irrigation control system. Meteorological data from weather forecasting services allow for forecasting evapotranspiration with sufficient accuracy on a "sliding" interval (Kovalchuk V. et al., 2019). The combination of instrumental methods with irrigation forecasting provides the most accurate determination of irrigation timing.

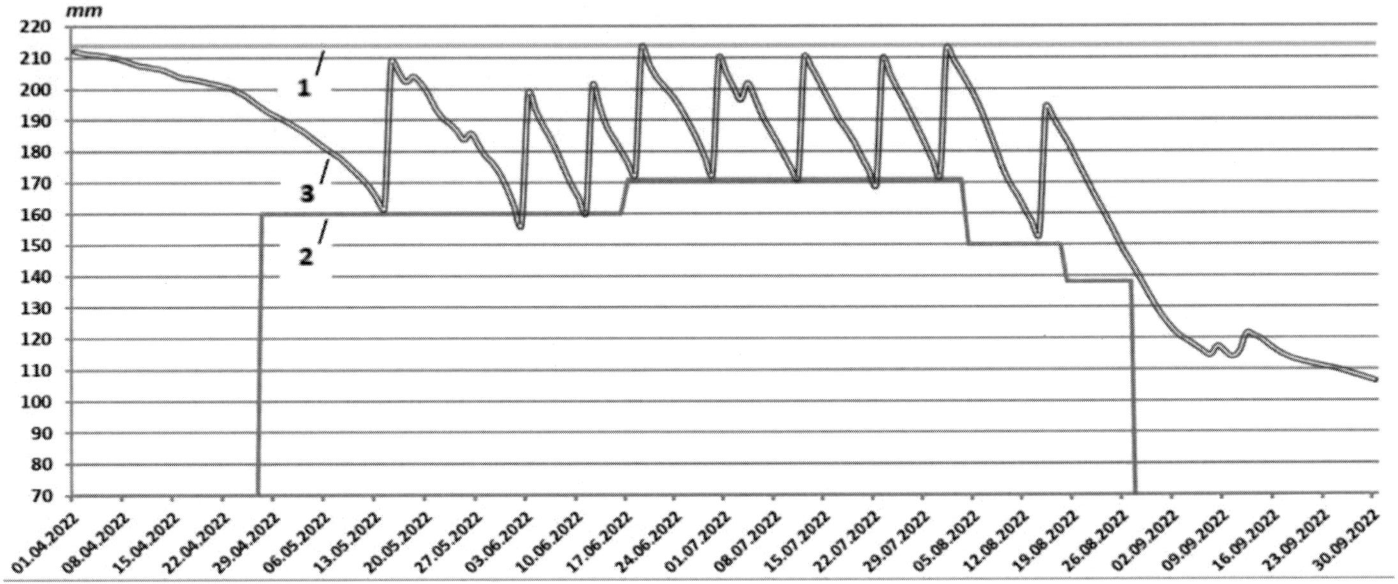
mm
220
210
200
190
180
170
160
150
140
130
120
110
100
90
80
70
1
3
2
01.04.2022
08.04.2022
15.04.2022
22.04.2022
29.04.2022
06.05.2022
13.05.2022
20.05.2022
27.05.2022
03.06.2022
10.06.2022
17.06.2022
24.06.2022
01.07.2022
08.07.2022
15.07.2022
22.07.2022
29.07.2022
05.08.2022
12.08.2022
19.08.2022
26.08.2022
02.09.2022
09.09.2022
16.09.2022
23.09.2022
30.09.2022

(a)

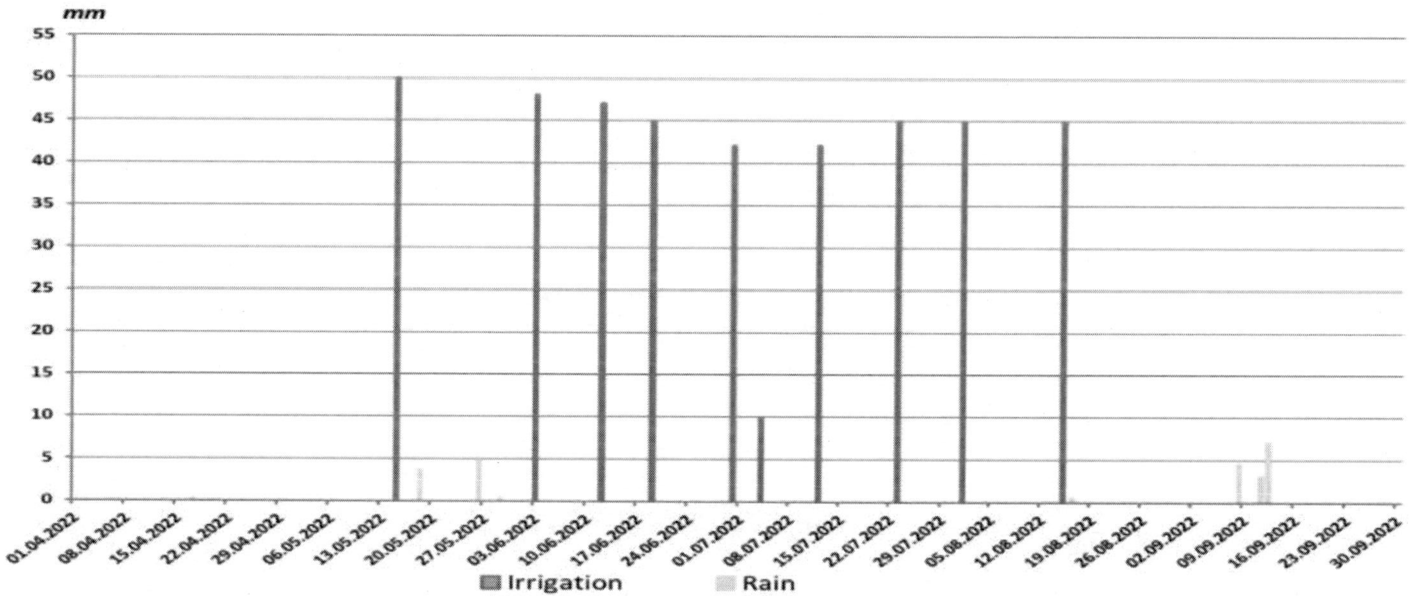

(b)

Figure 5. The graph of the dynamics of moisture content, irrigation and precipitation made by the results of operational irrigation control during the growing season of corn for grain; a: 1 – MMHC line; 2 – the line of the pre-irrigation soil moisture threshold differentiated by the stages of corn development; 3 – dynamics of soil moisture content; b: precipitation and irrigation during the growing season of corn for grain.

Alfalfa and corn for grain are essential indicators of the maximum water demand among crops in Ukraine. Alfalfa has the longest growing season. Corn is one of the highly profitable crops, so it was used to illustrate the practical introduction of the irrigation control algorithm. The biologically optimal irrigation regime for corn with a differentiated pre-irrigation soil moisture threshold, under optimal soil moisture conditions in the sowing period, showed water savings compared to the calculated rate.

The system approach, which includes an instrumental calculation algorithm, can be easily adapted to applying water-saving irrigation regimes. Applying water-saving irrigation regimes with a reduced irrigation rate and a pre-irrigation soil moisture threshold, differentiated by the stages of crop development, allows saving up to 30% of irrigation water compared to biologically optimal regimes (Gadzalo et al., 2019; Zhovtonog et al., 2015). At the same time, the risks of water infiltration into the deep layers of the soil are reduced, and the efficiency of using precipitation in conditions of uncertainty increases (Danlu et al., 2023).

An important area of resource conservation in global warming conditions is the selection of crop rotation structure. To reduce the total consumption of irrigation water, it is necessary to reduce the sown areas and - in some cases - to exclude moisture-loving crops such as corn for grain and soybeans from the cultivation on irrigated lands of Southern Ukraine. At the same time, it is recommended that the area of crops such as winter and spring cereals be increased in crop rotations. If in dry and extremely dry years (Table 2), the seasonal rate of irrigation for soybeans ranges from 2800-4200 m^3/ha, and it is 2900-4400 m^3 ha^{-1} for corn for grain, while the irrigation rate for winter wheat (taking into account autumn irrigation) is 1600-2200 m^3 ha^{-1}. This saves about 40% of irrigation water per unit area.

Conclusion

It was established that in climate change conditions, there is a need for a systematic approach to assessing the increase in crop water demand under irrigation. Therefore, it is expedient to propose an assessment method of the change in the deficit of general water balance for two-time intervals: 1960-1990 and 1991-2021. Calculations by this method showed a change in the water balance deficit over the years of different water supplies, which set the required level of resource conservation in irrigation. Depending on the rate of water balance deficit, models and decision-making methods for applying irrigation regimes were developed and applied for system irrigation management.

A system model and an instrumental calculation algorithm for operational irrigation control with forecasting on a "sliding" interval were developed for effective irrigation management in water shortage conditions. The instrumental approach uses soil moisture measurements for irrigation scheduling and provides continuous feedback on soil moisture status in the field. The proposed algorithms calculate and forecast irrigation terms and rates with the help of Internet weather services. This allows for the preparation of watering in advance and the effective use of precipitation.

Instrumental calculation algorithms are adapted to different irrigation regimes depending on the level of water conservation. If there is no limit to available water resources, it is advisable to use biologically optimal irrigation regimes. In conditions of significant saving of water resources, the algorithm adapts to the parameters of water-saving irrigation regimes. At the same time, economically justified irrigation rates should be established. It is necessary to introduce payment for water use at preferential tariffs when applying water-saving irrigation regimes. A significant water saving is achieved by introducing less moisture-loving crops into the cropping system.

Conflict of Interest

The authors have no competing interests to declare that are relevant to the content of this chapter.

Acknowledgments

The authors are grateful to Natalia Logunova for her help in preparing the materials for this chapter.

References

AccuWeather. 2023." *API Product Integration.*" Accessed November 06, 2023. https://www.accuweather.com/en/.

Adamenko, T. 2019. "Climate Change and Agriculture in Ukraine: What Should Farmers Know? German-Ukrainian Agro-political Dialogue." Dspace.organic-platform.org (423). https://dspace.organic-platform.org/xmlui/handle/data/423.

Allen, R. G., Luis S. Pereira, Dirk Raes, and Martin Smith. 1998. *Crop Evapotranspiration–guidelines for Computing Crop Water Requirements: FAO Irrigation and Drainage Paper 56*, Rome: FAO. Kindle eBook. https://www.fao.org/3/x0490e/x0490e00.htm.

Aruna, P., R. R. Raja, N. Rajeshwaran, M. Vignesh, and M. Vijayakumar. 2016. "Automatic Irrigation Control Using Wireless Sensor Network." *International Research Journal of Engineering and Technology (IRJET)*, 3(02): 1299-1302.

Barros, T. H. S., Asdrubal J. Farias, A. M. Almeida, Eusímio F. Fraga Jr, and Rubens D. Coelho. "Field Capacity Estimation from the Inflection Point of the Soil Moisture Retention Curve," in *IV INOVAGRI International Meeting, Fortaleza*, 2017, pp. 1-8. https://doi.org/10.7127/iv-inovagri-meeting-2017-res0250268.

Bianchi, A., Daniele Masseroni, Martin Thalheimer, Leonardo O. Medici, and Arianna Facchi. 2017. "Field Irrigation Management through Soil Water Potential Measurements: A Review." *Italian Journal of Agrometeorology* 22(2):25-38. https://doi.org/10.19199/2017.2.2038-5625.025.

Bouraima, A. K., WeiHua Zhang, and ChaoFu Wei. 2015. "Irrigation Water Requirements of Rice Using Cropwat Model in Northern Benin." *International Journal of Agricultural and Biological Engineering* 8(2):58-64. https://doi.org/10.3965/j.ijabe.20150802.1290.

Cao, J., Junwei Tan, Yuanlai Cui, Yufeng Luo. 2019. "Irrigation Scheduling of Paddy Rice Using Short-term Weather Forecast Data." *Agricultural Water Management* 213:714–723. https://doi.org/10.1016/j.agwat.2018.10.046.

Danlu, G., Quan J. Wang, Dongryeol Ryu, Qichun Yang, Peter Moller and Andrew W. Western. 2023. "An Analysis Framework to Evaluate Irrigation Decisions Using Short-term Ensemble Weather Forecasts." *Irrigation Science* 41:155–171. https://doi.org/10.1007/s00271-022-00807-w.

Ewaid, S. H., Ali A. Salwan, and Al-Ansari Nadhir. 2019. "Crop Water Requirements and Irrigation Schedules for Some Major Crops in Southern Iraq." *Water* 11(4):756. https://doi.org/10.3390/w11040756.

Faybishenko, B. 2012. "*Fuzzy-probabilistic Calculations of Evapotranspiration*." United States: Lawrence Berkeley National Laboratory. Berkeley, CA. https://doi.org/10.5772/18087. https://www.osti.gov/servlets/purl/1212435.

Gadzalo, Y., Mykhailo Romashchenko, Volodymyr Kovalchuk, Tetiana Matiash, and Olexandr Voitovich. "Using Smart Technologies in Irrigation Management," in: *International Commission on Irrigation and Drainage: 3rd World Irrigation Forum (WIF3), ICID Id: W.1.3.02.*, 2019, pp. 1–6. https://www.icid.org/wif3_bali_2019/wif3_1-3_2-min.pdf.

Goap, A., Deepak Sharma, A.K. Shukla, C. Rama Krishna. 2018. "An IoT Based Smart Irrigation Management System Using Machine Learning and Open Source Technologies." *Computers and Electronics in Agriculture* 155:41–49. https://doi.org/10.1016/j.compag.2018.09.040.

Kovalchuk, P., Tetiana Matyash, Volodymyr Kovalchuk, Olena Demchuk, Hanna Balykhina, Andriy Gerus, and Natalia Pendak. 2019. *System Modeling and Management of Water and Land Use: Monograph.* Kyiv: Agrarian Science. https://drive.google.com/file/d/16ZoZhIjAFHp-PK6AjPeVXH-gfVAsiKhY/view.

Kovalchuk, P., Viktoria Rozhko, Volodymyr Kovalchuk, Hanna Balykhina, and Olena Demchuk. 2021. "Crop Irrigation Demand Modelling Method within the Conditions of Climate Change in the Dnipro-Donbas Canal." *Lecture Notes on Data Engineering and Communications Technologies* 83:89-101. https://doi.org/10.1007/978-3-030-80472-5_8.

Kovalchuk, V. P. 2021. "Method of Obtaining Water-physical Properties of Soils by Fractions of Their Granulometric Composition." *Plant and Soil Science* 12(4):115-125. https://doi.org/10.31548/agr2021.04.115.

Kovalchuk, V., Oleksandr Voitovich, Dmytro Demchuk, and Olena Demchuk. 2021. "Development of Low-Cost Internet-of-Things (IoT) Networks for Field Air and Soil Monitoring within the Irrigation Control System." *Advances in Intelligent Systems and Computing (AISC)* 1247:86–96. https://doi.org/10.1007/978-3-030-55506-1_8.

Kovalchuk, V., Oleksandr Voitovich, Yarina Butenko, Svitlana Burykina, Vita Lukashuk, and Oleksandr Kovalchuk. 2023. "The Use of Weather Services, Remote and Field Methods to Justify the Temperature Indicators of the Development of Winter Wheat." *Bulletin of Agricultural Science* 5(842):71-80. https://doi.org/10.31073/agrovisnyk 202305-10.

Kovalchuk, V., Olena Demchuk, Dmytro Demchuk, and Oleksandr Voitovich. 2019. "Data Mining for a Model of Irrigation Control Using Weather Web-Services." *Advances in Intelligent Systems and Computing* 754:133-143. https://doi.org/10.1007/978-3-319-91008-6_14.

Liang, Z., Xiaochu Liu, Jianbin Xiong, and Jinrui Xiao. 2020. "Water Allocation and Integrative Management of Precision Irrigation: A Systematic Review." *Water 2020,* 12(11), 3135. http://dx.doi.org/10.3390/w12113135.

Lozoya, C., Carlos Mendoza, Alberto Aguilar, Armando Román, and Rodolfo Castelló. 2016. "Sensor-based Model Driven Control Strategy for Precision Irrigation." *Journal of Sensors*, Volume 2016, Article ID 9784071. http://dx.doi.org/10.1155/2016/9784071.

Madhusudhan, M. S., S. N. Vinay, J. C. Savitha, Gadad M. Nazeer, and M. N. Srikanth. 2021. "Crop Water and Net Irrigation Requirement of Major Crops Grown in Mandya City using Cropwat 8.0." *International Journal of Engineering Research & Technology (IJERT)* 10(06): 45-50. https://doi.org/10.17577/IJERTV10IS060022.

McKee, T. B., Nolan J. Doesken, and John Kleist, "The Relationship of Drought Frequency and Duration to Time Scales," in *Proceedings of the 8th Conference on Applied Climatology*, 1993, pp. 179-184.

Nath, S., N. Devakumar, Sitanshu S. Patra, and Abnishek Nanda. 2022. "Estimation of Crop Water Requirement and Irrigation Scheduling of Rice in Shivamogga District of Karnatakausing FAO CROPWAT." *Journal of Cereal Research* 14(Spl-2):18-26. http://doi.org/10.25174/2582- 2675/2022/125782.

OpenWeather. 2023." *Weather API."* Accessed November 06, 2023. https://openweathermap.org/.

Palmer, W. C. 1965. *Meteorological Drought*: *Research paper no.45*, Washington DC: U.S. Department of Commerce Weather Bureau. https://www.ncei.noaa.gov/monitoring-content/temp-and-precip/drought/docs/palmer.pdf.

Pandey, P. K., Parmendra P. Dabral, and Vanita Pandey. 2016. "Evaluation of Reference Evapotranspiration Methods for the Northeastern Region of India." *International Soil and Water Conservation Research* 4(1):52-63. https://doi.org/10.1016/j.iswcr.2016.02.003.

Payero, J. O., Ali Mirzakhani-Nafchi, Ahmad Khalilian, Xin Qiao, and Rebecca Davis. 2017. "Development of a Low-Cost Internet-of-Things (IoT) System for Monitoring Soil Water Potential Using Watermark 200SS Sensors." *Advances in Internet of Things* 7 (3):71-86. https://doi.org/10.4236/ait.2017.73005.

Perera, K. C., Andrew W. Western, Bandara Nawarathna, and Biju George.2014. "Forecasting Daily Reference Evapotranspiration for Australia Using Numerical Weather Prediction Outputs." *Agricultural and Forest Meteorology* 194:50-63. https://doi.org/10.1016/j.agrformet.2014.03.014.

Perera, K. C., Andrew W. Western, David E. Robertson, Biju George, and Bandara Nawarathna. 2016. "Ensemble Forecasting of Short-term System Scale Irrigation Demands Using Real-time Flow Data and Numerical Weather Predictions." *Water Resources Research* 52:4801-4822. https://doi.org/10.1002/2015WR018532.

Shtoiko, D. A., V. A. Pysarenko, O. S. Bychko, and L. I. Yelazhenko. 1977. "Estimated Methods for Determining Total Evaporation and Irrigation Time of Crops." *Zroshuvane Zemlerobstvo*, 21:3-8.

Sielianinov, G. T. 1966. *Agricultural World Climate Map.* Leningrad: Hydrometeorological Publishing House.

Smith, M., D. Kivumbi, and L. K. Heng. 2000. Use of the FAO CROPWAT Model in Deficit Irrigation Studies. In: *Deficit Irrigation Practices*. Rome: FAO. https://www.fao.org/3/Y3655E/Y3655E00.htm.

Solangi, G. Sh., Sabab A. Shah, Raied S. Alharbi, Sallahuddin Panhwar, Hareef A. Keerio, Junaid A. Memon, Tae-Woong Kim, and Ali D. Bughio. 2022. "Investigation of Irrigation Water Requirements for Major Crops Using CROPWAT Model Based on Climate Data." *Water* 14(16):2578. https://doi.org/10.3390/w14162578.

Thakur, D., Yugal Kumar, Arvind Kumar, and Pradeep K. Singh. 2019. "Applicability of Wireless Sensor Networks in Precision Agriculture: A Review." *Wireless Personal Communications* 107:471-512. https://doi.org/10.1007/s11277-019-06285-2.

Thalheimer, M. 2013. "A Low-cost Electronic Tensiometer System for Continuous Monitoring of Soil Water Potential." *Journal of Agricultural Engineering* 44(3), XLIV:e16. https://doi.org/10.4081/jae.2013.e16.

Toloknianenko, V. A. 1975. "Synthesis and Analysis of a Control System with a Sliding Optimization Interval." PhD diss., Kiev Polytechnic Institute.

Trnka, M., J. E. Olesen, K. C. Kersebau, A. O. Skjelvåg, J. Eitzinger, B. Seguin, and Z. Žalud. 2011. "Agroclimatic Conditions in Europe under Climate Change." *Global Change Biology* 17(7):2298-2318. https://doi.org/10.1111/j.1365-2486.2011.02396.x.

U.S. Department of Agriculture (USDA). Agricultural Research Service (ARS). 2024. *"ROSETTA Model (Rosetta Version 1.0)."* Accessed January 14, 2024. https://

www.ars.usda.gov/pacific-west-area/riverside-ca/agricultural-water-efficiency-and-salinity-research-unit/docs/model/rosetta-model/.

Ukrainian Academy of Agrarian Sciences. 2004. *Methodical Guidelines for the Operational Planning of Irrigation Regimes*. Kyiv: IHiM UAAN, IZPR UAAN. https://ias.pp.ua/pdf/Operational_planning_of_irrigation_2004.pdf.

Ushakov, E., and Oleg Vojtov. "Control with Optimization on a Sliding Interval by Continuous Processes," in *Proceedings of the International Conference Modern Problems of Radio Engineering, Telecommunications and Computer Science*, 2004, pp. 446-448.

Van Genuchten, M. Th. 1980. "A Closed-form Equation for Predicting the Hydraulic Conductivity of Unsaturated Soils." *Soil Science Society of America Journal* 44(5):892-898. https://doi.org/10.2136/sssaj1980.03615995004400050002x.

Vicente-Serrano, S. M., Santiago Beguería, and Juan I. López-Moreno. 2010. "A Multiscalar Drought Index Sensitive to Global Warming: the Standardized Precipitation Evapotranspiration Index." *Journal of Climate* 23:1696-1718. https://doi.org/10.1175/2009JCLI2909.1.

Wilson, L., Stacey New, Joseph Daron, and Nicola Golding. 2021. *Climate Change Impacts for Ukraine*. Exeter: Met Office. https://www.metoffice.gov.uk/binaries/content/assets/metofficegovuk/pdf/services/government/met-office_climate-change-impacts-for-ukraine_report_08dec2021_english.pdf.

Yang, M., Denghua Yan, YingdongYu, and Zhiyong Yang. 2016. "SPEI-based Spatiotemporal Analysis of Drought in Haihe River Basin from 1961 to 2010." *Advances in Meteorology* 2016:1-10. https://doi.org/10.1155/2016/7658015.

Yang, Y., Rensheng Chen, Chuntan Han, and Zhangwen Liu. 2021. "Evaluation of 18 Models for Calculating Potential Evapotranspiration in Different Climatic Zones of China." *Agricultural Water Management* 244:106545. https://doi.org/10.1016/j.agwat.2020.106545.

Zhovtonog, O., Larisa Filipenko, Taisia Demenkova, and Natalia Didenko. 2015. "The Use of the "GIS Polyv" Information System and the IRRIMET Module of the Internet Weather Station for Operational Planning of Irrigation during Sprinkling." *Tavrijskyi Scientific Bulletin* 92:159–165.

Biographical Sketches

Volodymyr P. Kovalchuk

Affiliation:

Department of Information Technologies and Marketing Innovations, Institute of Water Problems and Land Reclamation of NAAS.

Education:

1991-1996, Graduated from Heat Energy Faculty of National Technical University of Ukraine NTUU (KPI), Kyiv, Ukraine;

1997-2000, Postgraduate Student of Institute of Water Problems and Land Reclamation of NAAS, Ph. D. in Agricultural Land Reclamation (Engineering Sciences);

2016, Doctor of Engineering Sciences in Agricultural Land Reclamation (Engineering Sciences).

Business Address: 03022, Vasyl'kyvska str., 37, Kyiv, Ukraine.

Research and Professional Experience:

Sead research fellow position at the Institute of Water Problems and Land Reclamation of NAAS since 2016.

Professional Appointments:

Research toward the development of information technologies in irrigation agriculture, ecology, water and soil management, agricultural land reclamation.

Publications from the Last 3 Years:

[1] Romashchenko, M. I., Bohaienko, V. O., Matiash, T. V., Kovalchuk, V. P., & Krucheniuk, A. V. 2021. "Numerical simulation of irrigation scheduling using fractional Richards equation." *Irrigation Science, 39*(3), 385-396. doi: 10.1007/s00271-021-00725-3.

[2] Kovalchuk, P., Rozhko, V., Kovalchuk, V., Balykhina, H., and Demchuk, O. 2021. "Crop Irrigation Demand Modelling Method Within the Conditions of Climate Change in the Dnipro-Donbas Canal." *Lecture Notes on Data Engineering and Communications Technologies* 83:89-101. https://doi.org/10.1007/978-3-030-80472-5_8.

[3] Kovalchuk, V., Voitovich O., Demchuk D., and Demchuk O. 2021. "Development of Low-Cost Internet-of-Things (IoT) Networks for Field Air and Soil Monitoring Within the Irrigation Control System." *Advances in Intelligent Systems and Computing* 1247 AISC:86–96. https://doi.org/10.1007/978-3-030-55506-1_8.

[4] Kovalchuk, V., Voytovych, O., Butenko, Ya., Burykina, S., Lukashuk, V., and Kovalchuk, O. 2023. "The use of weather services, remote and field methods to justify the temperature indicators of the development of winter wheat." *Visnyk*

ahrarnoyi nauky 5(842):71-80. https://doi.org/10.31073/agrovisnyk202305-10. (in Ukrainian).

[5] Kovalchuk, P., Kovalchuk V., Balykhina H., Yatsiuk M., Demchuk O. and Nechai O., "Mathematical Modeling Methods of Evolutionary Dynamics of Wetland Ecosystems in a Changing Climate," *2023 IEEE 18th International Conference on Computer Science and Information Technologies (CSIT)*, Lviv, Ukraine, 2023, pp. 1-4, https://doi.org/10.1109/CSIT61576.2023.10324092.

[6] Kovalchuk, V., Voitovich, O., Kovalchuk, P., Demchuk, O. 2023. "Identification of Soil Water Potential Sensors Readings in an Irrigation Control System Using Internet-of-Things (IoT): Automatic Tensiometer and Watermark 200SS." *In: Hu, Z., Dychka, I., He, M. (eds) Advances in Computer Science for Engineering and Education VI. ICCSEEA 2023. Lecture Notes on Data Engineering and Communications Technologies*, vol 181. Springer, Cham. https://doi.org/10.1007/978-3-031-36118-0_54.

Hanna A. Balykhina

Affiliation:

Institute of Water Problems and Land Reclamation of the National Academy of Agrarian Sciences of Ukraine.

Education:

1996-2001 Graduated from the construction and hydromelioration Faculty of Kherson State Agrarian University;

2006-2007 MD degree in Hydro amelioration;

2013 pHD Agricultural Land Reclamation (Technical Science).

Business Address: 37 str. Vasylkivska, Kyiv, Ukraine.

Research and Professional Experience:

Senior researcher, ecology laboratory of the Institute of Water Problems and Land Reclamation NAAS since 2007.

Professional Appointments:

Senior researcher of the ecology laboratory.

Publications from the Last 3 Years:

[1] Kovalchuk, P., Rozhko, V., Kovalchuk, V., Balykhina, H., and Demchuk, O. 2021. "Crop Irrigation Demand Modelling Method Within the Conditions of Climate Change in the Dnipro-Donbas Canal." *Lecture Notes on Data Engineering and Communications Technologies* 83:89-101. https://doi.org/10.1007/978-3-030-80472-5_8.

[2] Kovalchuk, P., Kovalchuk V., Balykhina H., Yatsiuk M., Demchuk O. and Nechai O., "Mathematical Modeling Methods of Evolutionary Dynamics of Wetland Ecosystems in a Changing Climate," *2023 IEEE 18th International Conference on Computer Science and Information Technologies (CSIT)*, Lviv, Ukraine, 2023, pp. 1-4, https://doi.org/10.1109/CSIT61576.2023.10324092.

Pavlo I. Kovalchuk

Affiliation:

Department of Ecology and Nature Use, Institute of Water Problems and Land Reclamation of NAAS.

Education:

1963-1968, Graduated from Applied Mathematics Faculty of Ivan Franko Lviv State University, Lviv, Ukraine;

1971-1973, Postgraduate Student of Institute of Cybernetic of NAS of Ukraine, Ph. D. in Information Technology;

1991, Doctor of Engineering Sciences in Agricultural Land Reclamation (Engineering Sciences) and Information Technology;

Since 2001 Full Professor.

Business Address: 03022, Vasyl'kyvska str., 37, Kyiv, Ukraine.

Research and Professional Experience:

Head research fellow position at the Institute of Water Problems and Land Reclamation of NAAS since 2001.

Professional Appointments:

Applied mathematics research, system analysis and management in ecology, water quality, irrigation agriculture.

Editorial team member of International Journal "Land Reclamation and Water Management."

Publications from the Last 3 Years:

[1] Kovalchuk, P., Rozhko, V., Kovalchuk, V., Balykhina, H., and Demchuk, O. 2021. "Crop Irrigation Demand Modelling Method Within the Conditions of Climate Change in the Dnipro-Donbas Canal." *Lecture Notes on Data Engineering and Communications Technologies* 83:89-101. https://doi.org/10.1007/978-3-030-80472-5_8.

[2] Kovalchuk P., Kovalchuk V., Balykhina H., Yatsiuk M., Demchuk O. and Nechai O., "Mathematical Modeling Methods of Evolutionary Dynamics of Wetland Ecosystems in a Changing Climate," *2023 IEEE 18th International Conference on Computer Science and Information Technologies (CSIT)*, Lviv, Ukraine, 2023, pp. 1-4, https://doi.org/10.1109/CSIT61576.2023.10324092.

[3] Kovalchuk, V., Voitovich, O., Kovalchuk, P., Demchuk, O. 2023. "Identification of Soil Water Potential Sensors Readings in an Irrigation Control System Using Internet-of-Things (IoT): Automatic Tensiometer and Watermark 200SS." *In: Hu, Z., Dychka, I., He, M. (eds) Advances in Computer Science for Engineering and Education VI. ICCSEEA 2023. Lecture Notes on Data Engineering and Communications Technologies*, vol 181. Springer, Cham. https://doi.org/10.1007/978-3-031-36118-0_54.

Mykhailo V. Yatsiuk

Affiliation:

Institute of Water Problems and Land Reclamation of the National Academy of Agrarian Sciences of Ukraine.

Education:

Hydrologist- chemical hydrologist,
Kyiv National University n.a. Taras Shevchenko, 1990-1995
PhD in Geographical Sciences in the field of Water Safety
Kyiv National University n.a. Taras Shevchenko, 1995-1998

02/2022–till now Director of Institute of Water Problems and Land Reclamation of NAAS, Ukraine.

Business Address: 37 str. Vasylkivska, Kyiv, Ukraine.

Professional Appointments:

Director of the Institute of Water Problems and Land Reclamation NAAS since 2022.

Publications from the Last 3 Years:

[1] Yatsyuk, M., Sydorenko O., Tsvetova O., Turaieva O. Features of the Hydrological Regime of Shatsk Water Bodies of the Volyn Region. *Handbook of Research on Improving the Natural and Ecological Conditions of the Polesie Zone: monograph*. IGI Global. Hershey, PA, USA, 2023. 402-416 p. doi: 10.4018/978-1-6684-8248-3

[2] Kovalchuk, P., Kovalchuk V., Balykhina H., Yatsiuk M., Demchuk O. and Nechai O., "Mathematical Modeling Methods of Evolutionary Dynamics of Wetland Ecosystems in a Changing Climate," 2023 *IEEE 18th International Conference on Computer Science and Information Technologies (CSIT),* Lviv, Ukraine, 2023, pp. 1-4, https://doi.org/10.1109/CSIT61576.2023.10324092.

[3] Yatsiuk, M. V., Y. B. Mosiichuk; Ye. M. Matselyuk; A. B. Mosiichuk Analysis of the Hydrochemical Regime of the Dnipro Reservoirs Interagency thematic scientific collection *"Land Reclamation and Water Management"* 2023-07-02 doi: 10.31073/mivg202301-354

[4] Fedun, I., Kudyrko L., Yatsyuk M., Shnyrkov O., Bey R., Syniuchenko A. Economic and Political Challenges of Development in Ukraine Industry 4.0. In: Alareeni, B., Hamdan, A., Khamis, R., Khoury, R. E. (eds) Digitalisation: Opportunities and Challenges for Business. *Lecture Notes in Networks and Systems*, vol 620. Springer, Cham. pp. 453-467. https://doi.org/10.1007/978-3-031-26953-0_42.

Oleksandr P. Voitovych

Affiliation:

Institute of Water Problems and Land Reclamation, Department of ecology and nature use.

Education:

1999-2004 Kherson National Technical University, faculty Cybernetics
2016-2020 Institute of Water Problems and Land Reclamation, Ph. D. in Agronomy.

Research and Professional Experience:

Senior Research Fellow at Institute of Water Problems and Land Reclamation since 2023.

Professional Appointments:

Editor of a scientific journal "Land reclamation and water management."

Publications from the Last 3 Years:

[1] Kovalchuk, V., Voytovych, O., Butenko, Ya., Burykina, S., Lukashuk, V., and Kovalchuk, O. 2023. "The use of weather services, remote and field methods to justify the temperature indicators of the development of winter wheat." *Visnyk ahrarnoyi nauky* 5(842):71-80. https://doi.org/10.31073/agrovisnyk202305-10. (in Ukrainian).

[2] Kovalchuk, V., Voitovich, O., Kovalchuk, P., Demchuk, O. 2023. "Identification of Soil Water Potential Sensors Readings in an Irrigation Control System Using Internet-of-Things (IoT): Automatic Tensiometer and Watermark 200SS." *In: Hu, Z., Dychka, I., He, M. (eds) Advances in Computer Science for Engineering and Education VI. ICCSEEA 2023. Lecture Notes on Data Engineering and Communications Technologies*, vol 181. Springer, Cham. https://doi.org/10.1007/978-3-031-36118-0_54.

[3] Kovalchuk V., Voitovich O., Demchuk D., and Demchuk O. 2021. "Development of Low-Cost Internet-of-Things (IoT) Networks for Field Air and Soil Monitoring Within the Irrigation Control System." *Advances in Intelligent Systems and Computing* 1247 AISC:86–96. https://doi.org/10.1007/978-3-030-55506-1_8.

Dmytro O. Demchuk

Affiliation:

Institute of Water Problems and Land Reclamation of NAAS.

Education:

2014-2019 National Technical University of Ukraine "Igor Sikorsky Kyiv Polytechnic Institute," master's degree at Computer Engineering.

Business Address:

37 str. Vasylkivska, Kyiv, Ukraine.

Research and Professional Experience:

Ukrainian Patent for Utility Model UA132271, https://sis.ukrpatent.org/en/search/detail/1223767/.

Professional Appointments:

Postgraduate at Institute of Water Problems and Land Reclamation of NAAS.

Publications from the Last 3 Years:

Kovalchuk V., Voitovich O., Demchuk D., and Demchuk O. 2021. "Development of Low-Cost Internet-of-Things (IoT) Networks for Field Air and Soil Monitoring Within the Irrigation Control System." *Advances in Intelligent Systems and Computing* 1247 AISC:86–96. https://doi.org/10.1007/978-3-030-55506-1_8.

Chapter 3

Cactus Pear Irrigation Management and Production in Arid Areas: From Farm to Fork

Maha Elbana[1], PhD
Mahmoud Elwakeel[2,*], PhD
and Mohamed Rashad[3], PhD
[1]Department of Soil and Water Sciences, Faculty of Agriculture, Beni-Suef University, Beni Suef, Egypt
[2]Department of Food Sciences, Faculty of Agriculture, Beni-Suef University, Beni Suef, Egypt
[3]Land and Water Technologies Department, Arid Land Cultivation Research Institute, City of Scientific Research and Technological Applications (SRTA-City), New Borg Elarab, Alexandria, Egypt

Abstract

Water is essential for agricultural productivity and food security, especially in arid regions like Egypt. These areas face significant challenges related to water scarcity, developmental strides, and the sustainability of their agricultural domain. Furthermore, the encroachment of urban expansion on arable land presents considerable threats to alimentary stability and ecological integrity. An analysis of the Nile River basin's water budget indicates that the water consumption due to actual evapotranspiration exceeds the input from precipitation, primarily due to evapotranspiration from native flora that does not contribute to agricultural yield. Initiatives to supplant these species with

* Corresponding Author's Email: mahmoud.elwakeel@agr.bsu.edu.eg

In: Irrigation Management
Editors: Maha Elbana and Youssef Brouziyne
ISBN: 979-8-89530-318-4

other native crops with positive value to agriculture production to the Nile basin's biotope may reduce the excessive water expenditure through evapotranspiration, which is thus pivotal in establishing innovative agri-food systems. However, such strategic ecological replacements necessitate careful deliberation to avoid detrimental impacts on the biotic community or the already limited water resources. In this context, attention has been concentrated on marginal lands, irrespective of their native vegetation status, to restore the ecosystem and diminish the non-beneficial water loss through actual evapotranspiration or evaporation. Egypt's marginal zones are typically afflicted by severe water deficits, unpredictable and sparse rainfall, high temperatures, subpar soil fertility, and land degradation. Therefore, any vegetative species introduced into these inhospitable terrains must exhibit resilience to such harsh conditions and not vie for resources with the region's staple crops, thus preventing further stress on the scarce water reserves.
Opuntia ficus-indica, commonly called the cactus pear, is increasingly cultivated within the fragile ecosystem's characteristic of marginal and arid regions. This species demonstrates a remarkable capacity for water conservation, sequestering significant quantities within its cellular structure. Additionally, the plant's root system is adept at rapid and voluminous water uptake. These traits confer a distinct advantage in regions such as Egypt. The ensuing discourse will comprehensively examine the propagation of *Opuntia ficus-indica* in Egypt, beginning with its agricultural techniques and the advantages it presents for cultivation in marginal terrains, progressing to its nutritional value and utilization as an additive in functional foods.

Keywords: *opuntia ficus-indica*, deficit irrigation, fruit quality, food processing, antioxidant, vitamin C

List of Abbreviations

ET_a	Actual evapotranspiration
kc	Crop coefficient
ET_o	Mean reference evapotranspiration
SMC	Soil moisture content
P	Precipitation
Θ_{fc}	Field capacity
WP	Water productivity
ANOVA	One-Way Analysis of Variance
SPSS	Statistical Package for the Social Sciences

D	Soil depth in centimeters
P_b	Bulk density
Θ_i	Percentage of soil moisture content
Ya	Actual harvested fruits
W_t	Total applied water
MP	Mega Pixel
ROI	Region of interest
CC	Canopy cover
pH	potential of hydrogen
FDA	Food and Drug Administration
Aw	Water activity
TSS	Total soluble solids
HPLC	High-performance liquid chromatography
TPC	Total phenolic contents
GAE	Gallic acid equivalents
E-122	Carmoisine.

Water Scarcity Crises in Arid Regions

Egypt's agricultural sector is responsible for consuming ~85% of the nation's accessible conventional water resources (Hossain and Da Silva, 2013; Negm, 2019). A significant portion of Egypt's landmass is desert and rangelands, encompassing an area of nearly 100 million hectares (Kumar et al., 2018), and is characterized by a pronounced deficiency in water supply. Cultivating crops that are resilient to the severe environmental conditions prevalent in the Mediterranean region, particularly on arid marginal lands, could substantially enhance food security. Nonetheless, such agricultural practices must not exacerbate the strain on the already limited water resources of these areas. The genus Opuntia has begun integrating into the agricultural framework (FAO and ICARDA, 2017; Kumar K. et al., 2018). Specifically, *Opuntia ficus-indica* is recognized as one of the economically pivotal species within the Cactaceae family on a global scale (Kiesling, 1998). This species has found favorable conditions for growth in various African nations, including Algeria, Morocco, Tunisia, South Africa, and Ethiopia (FAO and ICARDA, 2017). Within Egypt, *Opuntia ficus-indica* predominantly thrives in sandy soils (Abo-El-Ela et al., 2001), with notable cultivation in regions such as Giza, Menofyia, Ismailia, Sharkia, and Behira (Ammar et al., 2004). The agricultural footprint of *Opuntia ficus-indica*, commonly known as the cactus pear, has undergone a notable expansion within the Egyptian territory, escalating from an initial 650 hectares in 1994 to a substantial 6000 hectares by 2008 (Bakr, 2019). This remarkable growth can be primarily ascribed to the initiatives implemented by the Egyptian government, which were aimed at fostering the cultivation of cactus pear in arid community settings. These initiatives were part of a broader strategy designed to mitigate the impacts of climate change and simultaneously bolster women's participation in the agricultural sector within these localities (Najjar, 2018).

Extensive scholarly research has been undertaken within agriculture, focusing on Opuntia ficus-indica. These scholarly endeavors have predominantly explored the species' phylogenetic origins, geographical dispersion, and environmental interactions (FAO and ICARDA, 2017; Kiesling, 1998; Kumar K. et al., 2018). Furthermore, specific analyses have scrutinized the socio-economic impacts of cultivating cactus pear on marginal terrains, particularly emphasising the augmented involvement of women in agrarian activities (Najjar, 2018). Authoritative manuals detailing the premier agronomic practices for *Opuntia ficus-indica* cultivation have been disseminated (CAAES, 2012; FAO and ICARDA, 2017; Liguori and Inglese, 2015), as well as its application in animal husbandry as feed (Bakr, 2019). A botanical assessment by (Khalafalla et al., 2016) cataloged 71 species cohabiting with *Opuntia ficus-indica*. Investigations by (Ammar et al., 2004) concentrated on the foremost pathogens responsible for the deterioration of the cladodes and fruits of *Opuntia ficus-indica* in Egypt, including *Fusarium solani*, *Alternaria alternata*, and *Botryodiplodia theobromae* as the primary agents. Regarding hydric resource management in agriculture, (Snyman, 2004) noted that the majority of *Opuntia ficus-indica*'s root biomass is initially situated within the uppermost 0.1 m of the soil stratum, which extends to an approximate depth of 1.8 m following the inaugural growth cycle. This observation concurs with the findings of (North and Nobel, 1992), who reported that the root system of the species can reach depths ranging from 4 to 8 m. Nonetheless, (Snyman, 2004) and (FAO and ICARDA, 2017) assert that the effective rooting zone of *Opuntia ficus-indica* does not exceed a depth of 0.30 m under ideal conditions.

Consoli, Inglese, and Inglese 2013, evaluated actual evapotranspiration (ET_a) and the crop coefficient (kc) for *Opuntia ficus-indica* through micrometeorological methodology. The research revealed that the average daily ET_a and kc were quantified at 2.5 mm and 0.40, in contrast to a mean reference evapotranspiration (ET_o) of 5.0 mm. Various techniques are employed to ascertain crop ET_a and its association with soil moisture content

(SMC). (Jung et al., 2010) integrated meteorological data with remote sensing of ET_a and SMC indicators, discovering a detrimental effect on global ET_a when soil moisture was limited. Duffková, Zajíček, and Nováková 2011, utilized the energy balance equation to estimate ET_a for various crops, taking into account different soil types, gradients, and properties and observed that soil characteristics markedly affect ET_a, especially during crop maturation and drought conditions, with ET_a being significantly reduced in coarse-textured soils as opposed to finer-textured soils. In a three-year study from 2010 to 2013, (Graf et al., 2014) estimated SMC through wavelet coherence analysis, employing a soil-water balance model that included precipitation (P), ET_a, ET_o, and runoff. Their findings indicated a positive correlation between the ET_a/ET_o ratio and the available water content in the soil profile.

Filgueiras et al., 2020, integrated regression algorithms with remotely sensed vegetation indices pertinent to maize cultivation to forecast ET_a and SMC. This endeavor is aimed at augmenting the efficacy of economical irrigation methodologies. The resultant models corroborated the utility of vegetation indices in refining the precision of ET_a and SMC estimations and their computational representations. In a parallel study, (Rahmati et al., 2020) employed a lysimeter for the quantification of ET_a within grassland ecosystems, concurrently examining its association with SMC through a soil water balance methodology, encompassing data from 2012 to 2018. Their research accentuated a marked interrelation between ET_a and SMC, particularly more pronounced within arid soils than those with higher moisture content. Conversely, (Seneviratne et al., 2010) in their scholarly review, drew attention to the significant ambiguities concerning the degree to which SMC can influence ET_a. Subsequent inquiries by (Koster et al., 2004) and (Seneviratne et al., 2010) posited that the influence of SMC on ET_a is substantial upon surpassing a specific threshold, beneath which vegetation encounters constraints in ET_a processes. They identified three principal soil moisture conditions: (i) the Wet regime, delineated by SMC levels exceeding the pivotal threshold; (ii) the Dry regime, defined by SMC descending beneath the wilting point; and (iii) the Transitional regime, where SMC is intermediate between the Wet and Dry states. Notably, within the Transitional regime, the most significant impacts of SMC on ET_a were discerned, with variances in SMC precipitating commensurate alterations in ET_a. Nonetheless, they observed that the Wet and Dry regimes exerted minimal influence on ET_a dynamics.

Empirical evidence has substantiated the resilience of *Opuntia ficus-indica* in withstanding the severe aridity characteristic of both arid and semi-arid regions. (De Souza et al., 2021) conducted a longitudinal study over two years (October 2015-August 2017), deliberately eschewing any form of irrigation or fertilization to ascertain the effects of protracted drought conditions on the 'Gigante' cultivar's yield and viability. The experimental design incorporated six divergent treatments: (i) a complete omission of fertilization and irrigation, (ii) exclusion of fertilization coupled with a weekly provision of 0.6 L of water per plant, (iii) a similar setup to the preceding but with an increased water allocation of 1.2 L, (iv) the arrangement mentioned above with bi-weekly water distribution, (v) integration of organic fertilization with 1.2 L of water per plant every week, and (vi) organic fertilization in the absence of irrigation. Results from this study revealed that the administration of 0.6 L of water per plant weekly markedly augmented the productivity of *Opuntia ficus-indica*, outperforming those reliant solely on natural precipitation and concurrently bolstering survival rates. However, there remains a notable shortage of research dedicated to quantifying agronomic water management indices for *Opuntia ficus-indica*, SMC, kc, and ET_a. These metrics are imperative for accurately compiling crop water requisites and optimizing water productivity at the agronomic scale. Therefore, a detailed study is presented in Experiment 1 to elucidate the ramifications of different irrigation regimens on the water management parameters of *Opuntia ficus-indica* (ET_a, kc, SMC), crop water productivity, and fruit quality (Elbana et al., 2020; Elbana, Elwakeel, and Rashad 2020).

Experiment No. 1

The experiment was executed in sandy loam soil characterized by a composition of 65.3% sand and 18.7% clay, with a bulk density measuring 1.56 g cm^{-3} (Elbana et al., 2020). SMC at saturation, field capacity, and permanent wilting point were quantified as 42.8%, 22.7%, and 13.4% by weight, respectively. The experiment is conducted in northwest Egypt under the Mediterranean arid climate typology. Meteorological data pertinent to the study's duration was acquired from the proximate New Borg Elarab Airport weather facility, with the data accessible via The Weather Company (©TWC Product and Technology LLC 2014, 2020). The mean minimum and maximum temperatures recorded were 15.7°C and 25.8°C, respectively. The total precipitation accumulated was 213.3 mm in 2019 and 256.0 mm until November 2020. Precipitation predominantly occurred during the winter and autumn months, from October to March. Conversely, the summer and spring months were characterized by arid conditions and elevated temperatures, with an average precipitation of zero mm.

The daily ET_o values were deduced employing the FAO Penman-Monteith formula, adhering to the methodologies delineated by (Allen et al., 1998), and computed utilizing the ET_o Calculator software (Version 3.2, ©FAO 2012). They exhibited significant seasonal fluctuation, with the lowest rate recorded at 0.8 mm day^{-1} during the colder winter months and peaking at 6.0 mm day^{-1} amidst the warmer summer. It is observed that the ET_o rates during the summer season, with an approximate average of 4.0 mm day^{-1}, are consistently higher in comparison to the winter season, which maintains an average near 1.5 mm day^{-1}. These variations are critical for understanding seasonal water requirements for agricultural planning and irrigation scheduling.

The *Opuntia ficus-indica* was planted under two drip irrigation regimes i.e., 'A' is a complete irrigation regime and 'B' is a severe deficit irrigation regime. In Treatment 'A,' the irrigation frequency was established weekly, administering a volume of 7.2 cubic meters per hectare, following the guidelines proposed by the Egyptian Ministry of Agriculture and Land Reclamation (CAAES, 2012). Conversely, in Treatment 'B,' irrigation protocol was initiated whenever the soil moisture content descended beneath the threshold of 30% field capacity (Θ_{fc}), and the irrigation was sustained until the effective root zone was replenished to 40% of the Θ_{fc}. All plants received 7.2 m^3 ha^{-1} of water at the beginning of the experiment, and then treatments 'A' and 'B' were applied after that to ensure equal initial conditions. Each treatment was cultivated with three replicates and was separated by a minimum of 8 meters from each other to minimize the risk of lateral water transfer between the cultivated areas. Before planting, the cladodes were immersed in a copper oxychloride solution (1 g l^{-1}) to preemptively address potential infections, adhering to the guidelines set forth by (CAAES, 2012). Subsequently, the cladodes were stored in a ventilated, shaded place for four weeks. Planting commenced on March 24, 2019, with the cladodes being partially buried in the soil, ensuring one-third remained exposed and oriented east-west to maximize sunlight exposure throughout the day. The inter-furrow and intra-furrow spacing were maintained at 4 and 3 meters, respectively.

A composite fertilizer comprising 12 kg of organic compost, 0.5 kg of agricultural sulfur, 0.5 kg of ammonium sulfate, and 0.5 kg of superphosphate, was administered to each plant. This blend was deposited into a half-cubic meter cavity beneath the soil's surface at the inception of the study. The cultivation process was governed by three distinct irrigation schedules predicated on soil moisture depletion within the effective root zone. The (USDA, 1998) defines soil water depletion as the requisite volume of water necessary to elevate the soil moisture content to field capacity within the effective root zone. The cactus fruits were harvested twice on August 4, 2019, and September 27, 2020. The effect of each applied irrigation scheduling on *Opuntia ficus-indica* canopy cover development (CC), fruit yield, actual evapotranspiration (ET_a), kc, and water productivity (WP) was studied. The One-Way Analysis of Variance (ANOVA) methodology, implemented within the SPSS statistical software, facilitated the determination of statistically significant differences in the results derived from the tested irrigation regimes. The statistical analysis was conducted using a significant threshold of 0.05 ($p \leq 0.05$), ensuring that the results obtained were subject to a standard level of confidence commonly accepted in scientific research.

The SMC assessments were conducted weekly and before each irrigation event to determine ET_a, utilizing the approach delineated by (Israelsen OW and Hansen VE, 1979). The formula below illustrates the computation of ET_a, where 'D' represents the soil depth in centimeters, 'P_b' denotes the soil bulk density in grams per cubic centimeter, 'Θ_{fc}' is the percentage of SMC at field capacity, and 'Θ_i' is the percentage of SMC preceding irrigation (Michael, 2006).

$$ET_a = D \cdot P_b \cdot (\theta_{fc} - \theta_i)/100$$

The planted crop coefficient (Kc) was obtained by dividing ET_a by ET_o, and water productivity was obtained by dividing the actual harvested fruits (Y_a) by the total applied water (W_t) following (Molden et al., 2003).

The quantification of canopy cover (CC) has been systematically conducted through the application of digital image processing methodologies, as evidenced by multiple research endeavors (Korhonen and Heikkinen 2009; Lee and Lee 2011; Alivernini et al., 2018; Xiong et al., 2019). The current study employed Image Processing and Analysis in Java (ImageJ 1.52 p software, National Institutes of Health, USA) for the CC estimation from digital photographs. These images were captured using a 12 MP mobile camera positioned at a consistent elevation of 2 meters, with a frequency of once per week. Three photographic replicates were obtained for each variant of irrigation scheduling, and their mean value was calculated (Elbana et al., 2020). The procedural outline for processing the digital images via the ImageJ software is delineated in detail through the following steps:

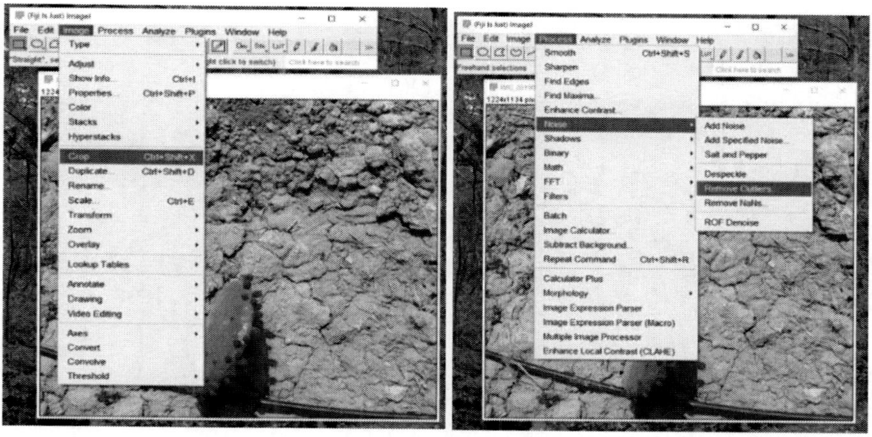

Figure 1. Step 1 for canopy cover processing using ImageJ software.

1. Digital images were centrally cropped as shown in figure 1 through the taps (Image > Crop) to mitigate potential errors arising from peripheral effects, and then noise and outliers were removed (Process > Noise > Remove Outliers).
2. Convert the image to Binary or in Black and white as shown in Figure 2 (Process>Binary>Make Binary)

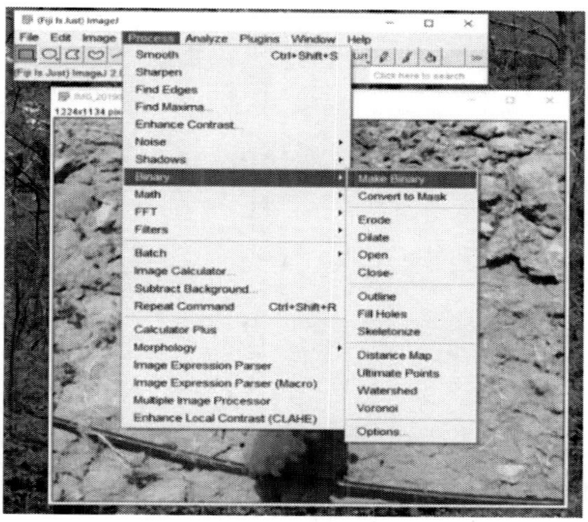

Figure 2. Step 2 for canopy cover processing using ImageJ software.

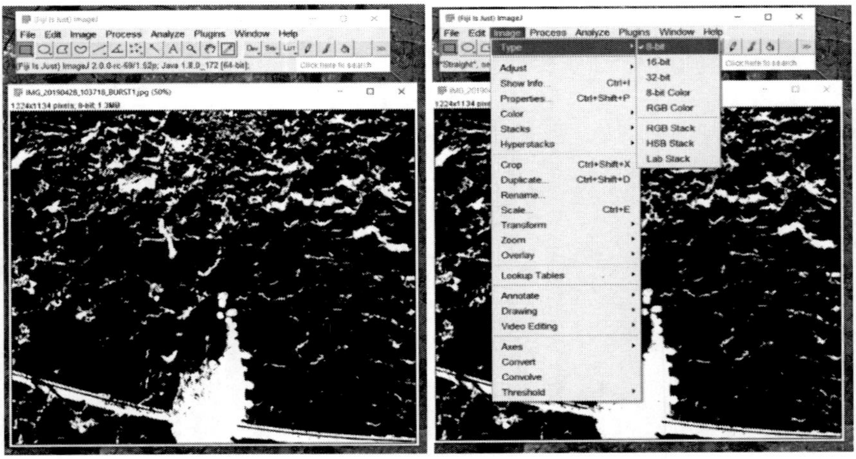

Figure 3. Step 3 for canopy cover processing using ImageJ software.

3. Make sure the image is 8-bit as demonstrated in Figure 3 (Image-Type-8 bit)
4. Adjust the image using a threshold for the best possible separation of canopy and background as shown in Figure 4 (Image> Adjust>Threshold), and uncheck the "Dark background" box.

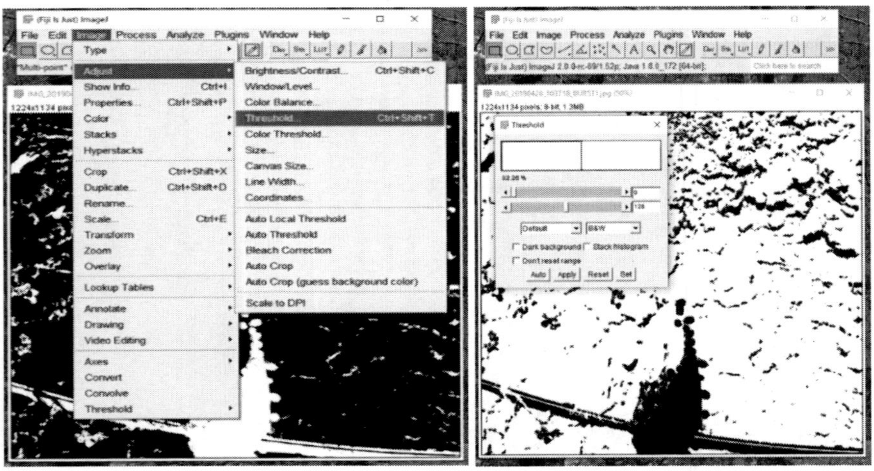

Figure 4. Step 4 for canopy cover processing using ImageJ software.

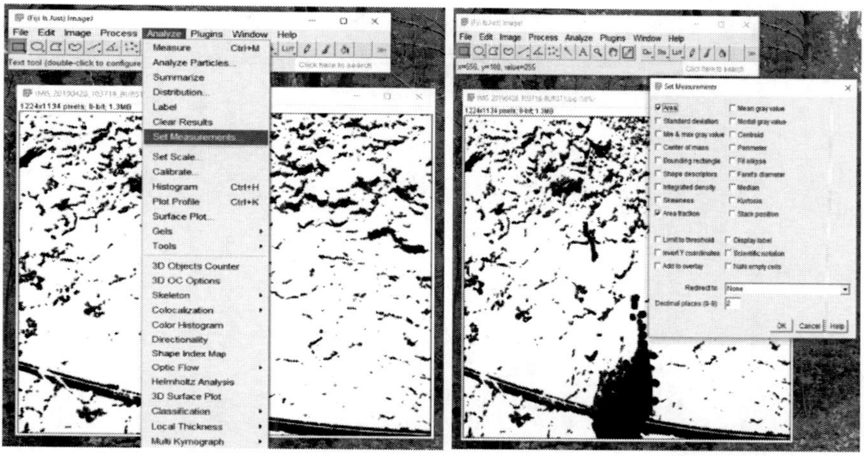

Figure 5. Step 5 for canopy cover processing using ImageJ software.

5. Set requested measurements as it could be seen in Figure 5 (Analyze>Set Measurements>Area Fraction)
6. Select the rectangle around the canopy to determine the region of interest as demonstrated in Figure 6 to determine the canopy cover, then add the region of interest (ROI) to the image analysis (Analyze>Tools>ROI Manager>Add[t]).

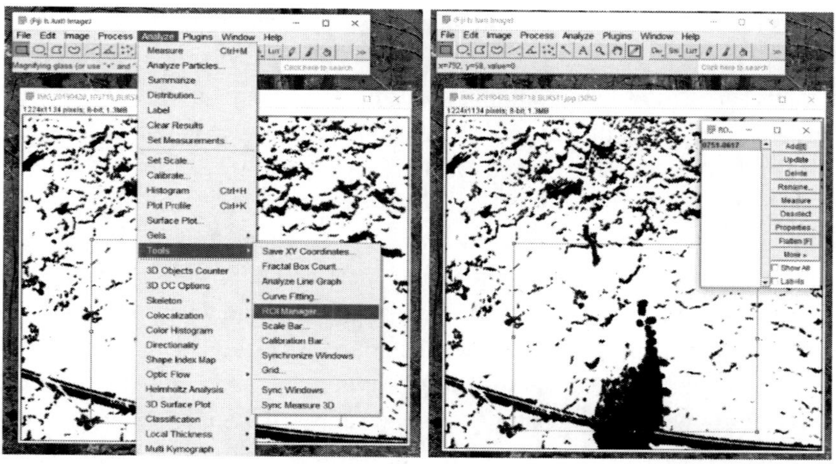

Figure 6. Step 6 for canopy cover processing using ImageJ software.

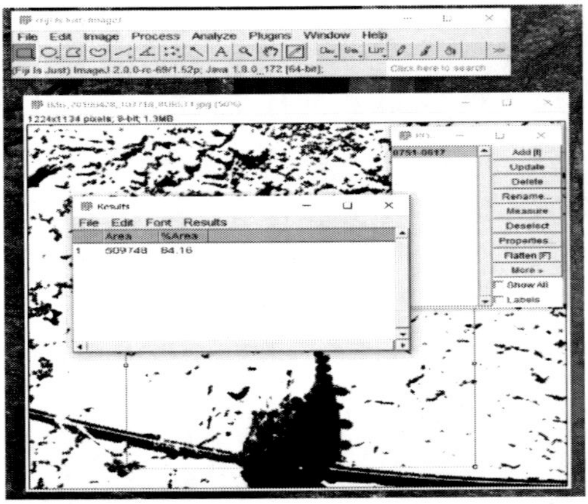

Figure 7. Step 7 for canopy cover processing using ImageJ software.

7. Select the added area in the ROI manager and click "Measure" to determine the percentage area of the white pixels as illustrated in Figure 7. Canopy cover (CC) is obtained by subtracting the % of white pixels from 100.

Table 1 delineates the mean and standard error for various parameters such as SMC on a weight basis, CC, fruit yield, WP, ET_a, and kc for *Opuntia ficus-indica* throughout the 2019-2020 experimental period. Statistical analysis revealed a significantly higher mean SMC in treatment 'A' ($P \leq 0.05$) compared to treatment 'B.' This discrepancy is likely attributable to the greater volume of irrigation water administered in 'A.' Specifically, 'A' received a cumulative irrigation of 554 $m^3\ ha^{-1}$, attributed to a weekly irrigation rate of 7.2 $m^3\ ha^{-1}$, whereas 'B' received 12.54 $m^3\ ha^{-1}$ due to irrigation being applied only when SMC fell below 30% of Θ_{fc}. The irrigation schedule for 'B' predominantly targeted the summer season, characterized by elevated temperatures, reduced precipitation, and increased ET_o. During winter, the precipitation contributed to maintaining the SMC within the targeted range in the effective root zone. This maintenance of SMC may also be facilitated by the rapid water uptake capabilities of *Opuntia ficus-indica*, the presence of a moderately impermeable coating over the roots - particularly the fine roots within the effective root zone - and the plant's ability to modulate shoot transpiration as necessary (FAO and ICARDA, 2017; Passioura, 1988).

Table 1. The mean and standard error for Opuntia ficus-indica fruit yield, water productivity (WP), canopy cover (CC), soil moisture content (SMC), actual evapotranspiration (ETa), and crop coefficient (Kc) during the experiment

Irrigation regime	Yield Kg ha^{-1}	WP Kg m^{-3}	CC %	SMC %	ET_a mm day^{-1}	Kc dimensionless
A	342	0.62	13.63 ± 1.28^a	12.78 ± 0.53^a	3.84 ± 0.20^b	0.19 ± 0.01^a
B	227	18.13	10.22 ± 1.13^a	10.47 ± 0.58^b	4.80 ± 0.22^a	0.24 ± 0.22^a

[*] Distinct lowercase letters a and b denote a statistically significant difference at the 0.05 level across the applied irrigation regimes.

Aside from the apparent substantial variance in SMC between the applied irrigation regimes, there was an absence of a significant difference at $P \leq 0.05$ in the CC of *Opuntia ficus-indica* under the irrigation treatments 'A' and 'B.' This observation is corroborated by (De Souza et al., 2021), who reported no significant variation in the number of cladodes of *Opuntia ficus-indica* across four distinct irrigation regimes, ranging from the complete absence of irrigation over two consecutive years (2015-2017) to a maximum of 1.2 liters per plant per week, in the absence of fertilization during the experimental period. The estimated progression of CC for the plant throughout 2019 and 2020 is depicted in Figure 8. The noted reduction in CC values towards the end of September 2019 can be attributed to the pruning activity conducted. On September 17, 2019, after the initial harvest, the plants underwent a pruning process, which entailed retaining only two cladodes per parent cladode to enhance sunlight penetration within the canopy, thereby facilitating the growth of cladodes, as well as their flowering and fruit development, as described by (Liguori and Inglese, 2015).

The study revealed that the maximum percentage of CC area was 35% for 'A' and 28% for 'B,' suggesting that the plants were in a developing growth phase. Research by (Hassan et al., 2020) found that *Opuntia ficus-indica* cladodes fully expanded their surface area by the end of the initial growth season. The current investigation noted an augmentation in canopy cover during the second season, attributable to the emergence of new cladodes from the primary cladode. This phenomenon may elucidate the superior yield of 'A' (342 kg ha^{-1}) over 'B' (227 kg ha^{-1}), as a greater CC in 'A' correlated with an increased cladode count. Conversely, WP was minimal for the 'A' irrigation regime (0.62 kg m^{-3}), whereas the 'B' achieved the maximal WP (18.13 kg m^{-3}). WP represents the mass of yield, which, in this context, are fruits produced per cubic meter of water utilized. Despite 'A' attaining a higher yield by mass, it also incurred the greatest water consumption relative to 'B.' On the other hand, ET_a is influenced by vegetative growth (Field et al., 1998; Phocaides, 2007) and SMC (Koster et al., 2004; Seneviratne et al., 2010; Graf et al., 2014; Rahmati et al., 2020). The mean daily ET_a values were recorded as 3.84 and 4.80 mm day^{-1} for 'A' and 'B,' respectively. Analysis revealed no significant variance ($P \leq 0.05$) in the kc across the irrigation schedules, with values of 0.19 for 'A' and 0.24 for 'B.' These ET_a figures are notably greater than the 2.5 mm day^{-1} reported (Consoli et al., 2013), who also calculated a higher mean crop coefficient of 0.40 compared to this study. Although this research aligns with (Consoli et al., 2013) regarding the species studied, *Opuntia ficus-indica*, under Mediterranean climatic conditions, the microclimatic variables differed. Geographical location and microclimate conditions have been identified as significant factors influencing ET_a estimates, as reported by (Seneviratne et al., 2010). (Consoli et al., 2013) conducted their research in Italy, located north of the Mediterranean, where the average daily ET_o was 5 mm day^{-1}, in contrast to the 3 mm day^{-1} average ET_o in the current study (Elbana et al., 2020), potentially accounting for the discrepancies in ET_a estimations.

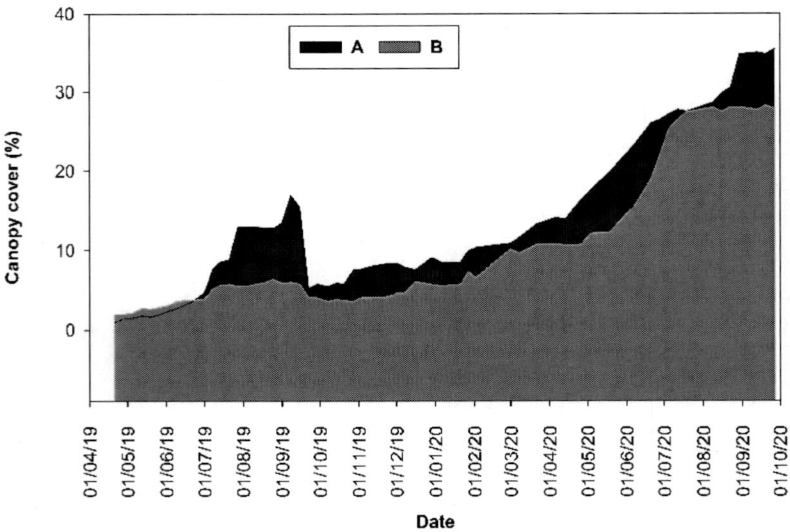

Figure 1. Estimated canopy cover development (%) during the 2019-2020 growing season for Opuntia ficus-indica under complete 'A' and deficit 'B' irrigation regimes.

Nutritional and Additive Values of Cactus Pear

The Opuntia species has experienced a surge in economic significance due to its nutritional and medicinal attributes (Piga, 2004). The fruit is composed of a substantial concentration of vitamin C, dietary fibers, and β-carotene (Albano et al., 2015), in addition to a suite of vital minerals such as calcium, phosphorus, magnesium, potassium, sodium, and iron (CAAES, 2012), coupled with a range of antioxidants (Sumaya-Martínez et al., 2011). Conventionally consumed fresh, the cactus pear is also processed into various products, including jams, juices, and dried snacks (Cassano et al., 2010), and is distinguished by its palatable, succulent interior and fragile exterior (Barbera et al., 1992). The fruit's intrinsic functional attributes are largely ascribed to its antioxidant capacity and substantial phenol and vitamin C composition (Sumaya-Martínez et al., 2011). It is acknowledged as a significant source of ascorbic acid and encompasses diverse compounds such as betalains, flavonoids, phenolic acids, and other bioactive constituents like polyphenols and specific pigments (Coria Cayupán et al., 2011), which confer nutraceutical and functional benefits (Cassano et al., 2010) and are purported to mitigate cardiovascular diseases (Jacob et al., 2008). Comparative analyses have revealed that the total antioxidant potential of the cactus pear markedly surpasses that of apples, pears, grapes, and tomatoes, and is on par with that of grapefruits and red grape raisins (Livrea and Tesoriere, 2006). In recent decades, the cactus pear has attracted increased attention due to its functional and nutritive qualities (Piga, 2004).

Betacyanins are composed of red pigments, while yellow to orange betaxanthins are present in purple plant varieties. The compound class known as betalains is recognized for its extensive pH stability range, antioxidative attributes, and potential contributory role in mitigating the progression of certain cardiovascular diseases (Sakuta, 2014). The confectionery product known as cakes, which enjoys widespread popularity across diverse age groups, undergoes both chemical and mechanical leavening processes (Rodríguez et al., 2006).

The fruit yielded from *Opuntia ficus-indica* is distributed through an extensive network of street vendors, with the fruit predominantly being consumed in its raw, unprocessed form. Despite this, there remains a discernible paucity of comprehensive data about the international commercial farming and production methodologies employed for the cactus pear (Saenz, 2000). The crop's adaptability is further demonstrated by its utilization in fabricating beverages, preserves, and medicinal products, among other commercial applications (FAO and ICARDA, 2017; Kumar K. et al., 2018; Saenz, 2000).

The United States Food and Drug Administration (FDA) has authorized the use of more than nine synthetic dye formulations, including three variants of red colorants. These artificial dyes are extensively utilized in a diverse spectrum of consumable products, encompassing items such as breakfast cereals, assorted sweets, and various frozen treats. Studies have established a correlation between the ingestion of synthetic dyes and the onset of inflammatory responses within biological systems, instigating the activation of the immune system, culminating in an elevated count of leukocytes within the bloodstream. Investigations have revealed that synthetic colorants may be implicated in the genesis of carcinogenic formations within the bladder and thyroid glands of rodent models, as well as causing perturbations in the immune cells distributed throughout the liver, spleen, and lymphatic tissues (Kobylewski and Jacobson, 2012).

The longevity and preservation of freshness in commercially available cupcakes are subject to considerable variation, influenced by factors including shelf-life and the type of ingredients utilized. The shelf-life of these confections typically spans from one to four weeks, occasionally extending further, contingent upon the quality of packaging, water activity levels (aw), and storage temperatures. Premium cupcakes are characterized by attributes such as increased volume, prolonged shelf-life, the firmness of the cake crumb, and uniformity in crumb structure. These qualities depend on an optimal combination of ingredients; meticulous batter mixing; the consistency of the batter before baking; and the heat-induced setting phase that follows. It has been observed that cakes with higher specific volumes are often associated with softer textures (Gomez-Conde et al., 2007).

The superiority of the final baked confection is fundamentally dependent on the quality of the constituent ingredients and their synergistic interactions. Within the domain of bakery products, cakes are categorized as air-leavened entities and are distinguished by specific hallmark characteristics when crafted from superior ingredients. These characteristics encompass an elevated volume, a homogenous crumb configuration, a soft texture, and a prolonged shelf-life. The ascension observed in cake products during baking can be ascribed to the entrapment of air vesicles within the batter, coupled with the complex fluid dynamics operative throughout the baking period. The ingredient composition of cakes is pivotal, exerting a considerable impact on both the structural integrity and the gustatory appeal. The integration of lipids is known to modify the structural attributes of the cake, affecting both the raw batter and the culminated baked well (Chaiya and Pongsawatmanit, 2011). The use of cactus pear pulp is noted for its capacity to augment coloration, impart flavor nuances, and contribute to the viscosity of the batter (Betancourt et al., 2017).

Cupcakes, categorized within the spectrum of confectionery products, frequently exhibit a visual similarity to muffins; nonetheless, they are characterized by a heightened sweetness and the absence of a cream-based garnish (Romjaun and Prakash, 2013). Contemporary shifts in dietary preferences have escalated the demand for pre-prepared consumables readily accessible in diminutive servings globally. The marketplace has witnessed an upsurge in the availability of products augmented with dietary fiber and antioxidants, catering to the escalating consumer inclination towards nutritious alternatives (Ajila et al., 2008). Amidst the burgeoning epidemic of obesity, which has precipitated a rise in the demographic afflicted with excess weight, these nutrient-dense foods have assumed a pivotal role in the regulation of body mass (Attia et al., 1993). Approximately 1,200 botanical species, renowned for their fibrous and antioxidative constituents, are becoming increasingly favored for their salutary benefits. These species are being incorporated as partial flour substitutes in the production of bakery items such as low-caloric bread and pastries, thereby bestowing a distinctive characteristic upon these comestibles. Nevertheless, the overutilization of specific additives has been associated with adverse impacts on the baking quality, manifesting as a granular consistency and a diminution in the size of the baked goods during the cooking process. Such additives impede the comprehensive hydration of the dough, consequently hindering its capacity to soften and amalgamate effectively within the blend (Brian et al., 1989).

The present investigation explained in experiment 2 (Elwakeel et al., 2021) was initiated with distinct objectives: (1) to elucidate and evaluate the impact of diverse irrigation methodologies, particularly under conditions of severe water scarcity, on the root zone's role in modulating fruit yield and water use efficiency; (2) to guarantee the production of fruit that adheres to commercial quality criteria; and (3) to aid in the restoration of ecosystems within arid territories by advocating for the cultivation of cactus pear utilizing minimal water resources. This study has examined the effect of betacyanins extracted from cactus pear fruits, contrasting their influence under different irrigation regimes, thereby identifying them as a novel natural betacyanin source.

Experiment No. 2

The fruits harvested under the two applied irrigation regimes described in Experiment 1 were subjected to being analyzed to conduct Experiment 2. The physical properties of cactus pear fruits were determined by measured diameter, length, and peel thickness by callipers while the weight and the pulp and thick peel were separated, and the weight of these fractions was noted following (Karababa et al., 2004). The fruit's shape index was determined according to the equation provided by (Mashope, 2007) as follows: fruit width/fruit length oblong = 0.045-0.550, elliptic = 0.560-0.600, ovoid = 0.700-0.790, and round = 0.800-0.890. The observed fruit's physical parameters are illustrated in Figure 9. The study revealed no significant variance in fruit weight between the standard irrigation regime 'A' and the water-deficient regime 'B,' although the pulp weight of 'B' (31.04 g) exceeded that of 'A' (26.33 g), both within the acceptable weight range for cactus pear fruits as established by (Makwana and Sabhadiya, 2022). Concurrently, the peel thickness of the fruits did not differ significantly across the irrigation treatments, maintaining a thickness of 6 mm or less, a preference among consumers for ease of handling, as noted by (Tsegay et al., 2018).

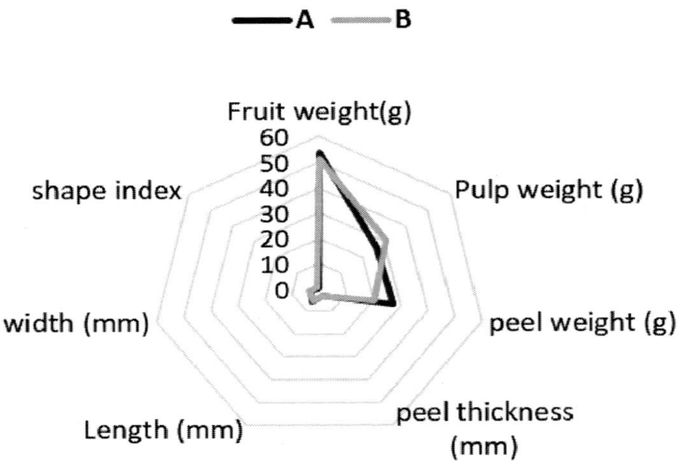

Figure 9. Opuntia ficus-indica fruit physical properties under complete (A) and deficit (B) irrigation regimes.

The irrigation treatments influenced the fruit shape index, indicating an elliptical shape for both 'A' and 'B' treatments. It has been documented that oval or barrel-shaped cactus pears are more manageable and sustain less stem-end damage during harvest compared to elongated fruits (da Silva et al., 2021). The concentration of betacyanin in the harvested fruits under the 'A' and 'B' irrigation regimes is shown in Figure 10, and the fruit pH and total soluble solids (TSS) content are shown in Figure 11. The betacyanins in cactus pear fruits were Determined using the HPLC device. Pure natural betanin was isolated from cactus pear fruit pulp, a rich source of betanin isomer extraction (Skalicky et al., 2020). Cactus pear pulp exhibited high betacyanin content, with aqueous extract samples showing relatively elevated betacyanin concentrations, imparting the fruit pulp with its distinctive deep red-violet hue. Notably, deficit irrigation correlated with increased water conservation, yet all treatments contained betanin as the primary betacyanin compound. The betacyanin levels in 'A' were determined to be 0.47 mg per 100 mg of dry weight in the fruit pulp, marginally higher than the 0.45 mg per 100 mg in 'B,' as reported by (Prieto-Santiago et al., 2020).

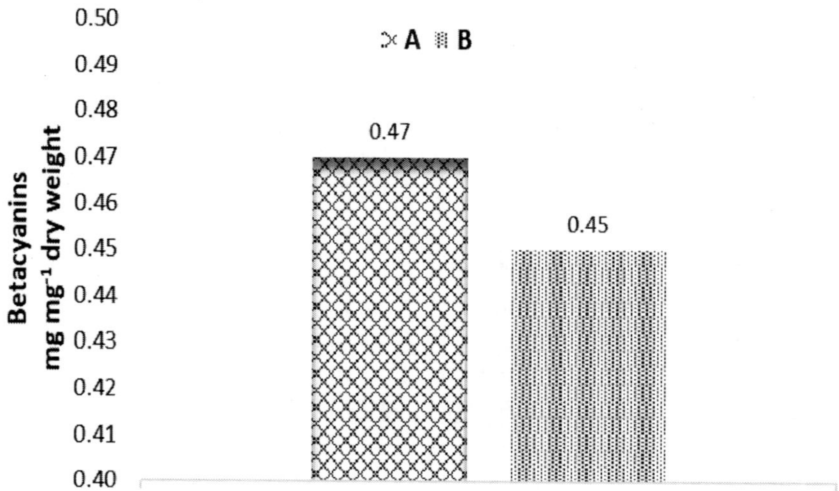

Figure 10. The concentration of betacyanins in Opuntia ficus-indica fruits cultivated under complete (A) and deficit (B) irrigation regimes.

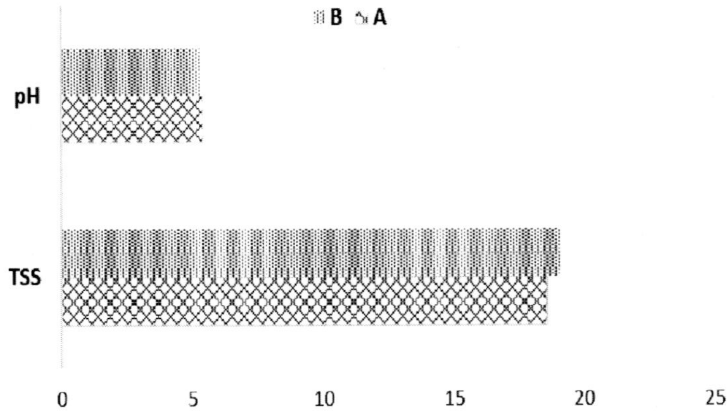

Figure 11. Total soluble solids and pH of the harvest Opuntia ficus-indica fruits under complete (A) and deficit (B) irrigation regimes.

The soluble solids content was consistent at approximately 18° Brix for all treatments, corroborating the results reported by (Du Toit et al., 2019; Sigwela et al., 2021). The *Opuntia spp.* has been identified as a significant source of ascorbic acid. The ascorbic acid content in the fruit pulp was determined according to (Dürüst et al., 1997) expressed in mg ascorbic acid per liter of pulp (mg AA L^{-1}). Comparable to many fruits and vegetables, the cactus fruit is an excellent natural source of vitamin C and can serve as a dietary supplement, particularly in arid regions where it is plentiful. The ascorbic acid content in the fruits of 'A' and 'B' was nearly identical, with values of 55.42 mg 100 g^{-1} and 56.04 mg 100 g^{-1}, respectively, as supported by (Du Toit et al., 2019). Furthermore, the pH levels of cactus pear pulp in all treatments were measured at around 5.3, consistent with the findings of (Chavez-Santoscoy et al., 2009).

The total phenolic contents (TPC) were determined using the Folin-Ciocalteau method (Bendhifi Zarroug et al., 2015). The TPC was expressed as gallic acid equivalents (GAE) and then compared to the gallic acid calibration curve. All values were expressed as mean (mg of gallic acid equivalents 100 ml^{-1}). The total flavonoid content in the pulp extracts was determined using spectrophotometry based on the formation of complex flavonoid-aluminum according to (Djeridane et al., 2006). The cactus pear, increasingly recognized for its health-promoting potential, exhibits a significant phenolic composition. These phenolic compounds, a varied class of plant-derived secondary metabolites, are linked to a spectrum of beneficial properties, including antioxidant, anti-inflammatory, and antimicrobial effects (Abou-Zaid, Ahmed, and Zedan, 2022). The Total Phenolic Content (TPC) in cactus pear fruit is quantified to be within the range of 50 to 300 mg of Gallic Acid Equivalents (GAE) per 100 grams of fresh weight, although this can fluctuate markedly across different varieties and environmental growth conditions. For instance, fruits cultivated under the 'A' irrigation regime demonstrated a TPC of 104.30 mg GAE 100 ml^{-1}, while those planted under the 'B' irrigation regime exhibited 95.37 mg GAE 100 ml^{-1}. These findings are in concordance with (Betancourt et al., 2017), who reported TPC in the pulp ranging from 11.6 to 10.8 g GAE kg^{-1}. The primary phenolic constituents identified in cactus pear include flavonoids such as isorhamnetin, quercetin, kaempferol, betalains like betanin, indicaxanthin, phenolic acids including gallic acid, ferulic acid, and p-coumaric acid. Notably, the phenolic profile is subject to variation contingent upon the cultivar, stage of maturity, and processing techniques (Palmeri et al., 2020). Analysis of the flavonoid content revealed concentrations of 1.44 and 1.41 ± 0.02 mg 100 ml^{-1} for 'A' and 'B' fruits, respectively, with 'A' presenting a slightly lower value. Literature indicates that cactus pear varieties are rich in flavonoids such as isorhamnetin, quercetin, and luteolin, all recognized for their antioxidant capabilities (Osuna-Martínez, Reyes-Esparza, and Rodríguez-Fragoso 2014; Zenteno-Ramírez et al., 2018).

Fruit Applications in Bakery Products

The consumption spectrum of bakery products has broadened, correlating with lifestyle shifts. Consequently, integrating natural additives like the red-violet pigment from Cactus pear juice has yielded positive outcomes in cake products, both technologically and nutritionally, presenting a viable substitute to synthetic colorants that may adversely affect public health (Scarano et al., 2022). Cupcakes, a subset of bakery items similar to muffins yet distinctively sweetened without cream topping, have adapted to contemporary dietary trends. The escalating consumer preference for convenience foods in small portions has been met with an increased production of such items. The market has concurrently seen a surge in products enriched with dietary fiber and antioxidants, responding to heightened consumer demand for these nutritional components (Bousbia et al., 2022). A combination of cactus pear juice with a confectionery matrix, such as cake, was investigated by substituting it for traditional ingredients at varying ratios of 20%, 40%, 60%, and 80%. This substitution was benchmarked against the addition of Carmoisine (E-122), a synthetic colorant as shown in Figure 12.

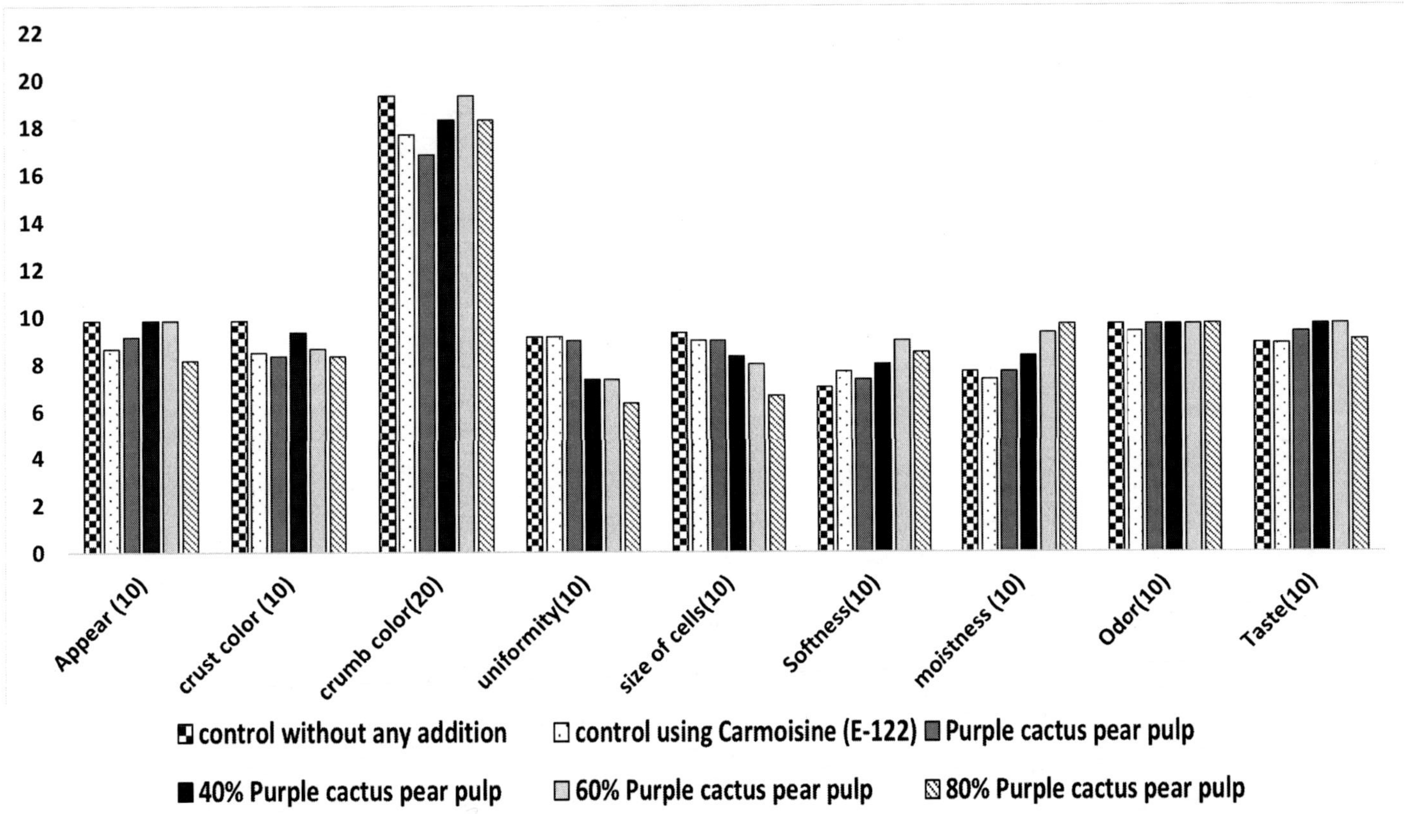

Figure 12. Total soluble solids and pH of the harvest *Opuntia ficus-indica* fruits under complete 'A' and deficit 'B' irrigation regimes.

Sensory analysis was conducted to evaluate attributes including appearance, crust and crumb color, cell uniformity, texture, moisture content, aroma, flavor, and overall acceptability by a panel of judges (Elwakeel et al., 2021). The sensory evaluation revealed that a 60% inclusion rate of cactus pear juice significantly enhanced all tested sensory properties, surpassing the performance of Carmoisine and other additives. The observed improvements are likely attributable to the intermolecular interactions between the fruit pulp, pectin, and other cake constituents. Consumer assessments of the cake's springiness - a measure of its ability to regain shape bles textural attribute, and the cake's overall quality, may be influenced by the hydrocolloids present in the cactus pear pulp (Gomez-Conde et al., 2007). Consequently, additing cactus pear pulp is posited to impact the springiness of the cake product beneficially.

Conclusion

Opuntia ficus-indica is an adequate plant for cultivation in arid marginal lands with severely limited water availability. It exhibited remarkable resilience under arid conditions, yielding fruit of considerable nutritional merit characterized by a rich array of antioxidants and betacyanins. The development of crop canopy cover and the crop coefficient remain unaltered by significant water shortages within the effective root zone. Furthermore, plant evapotranspiration rates were elevated when irrigation occurs in the summer compared to a weekly irrigation schedule. Conversely, a weekly irrigation regimen resulted in a greater fruit yield, measured in kilograms per hectare, than irrigation limited solely to the summer season. Incorporating cactus pear fruit into confectionery items like cupcakes enhances their nutritional profile. Substituting artificial colorants with cactus pear pulp in wheat flour formulations may produce cupcakes with improved nutritional value. Experimental cupcakes augmented with cactus pear pulp demonstrated enhanced flavor profiles, texture, and structural integrity. Nonetheless, additional research is necessary to optimize the sensory attributes of cupcakes fortified with cactus pear pulp.

Disclaimer

None.

References

Abo-El-Ela, A., Baiuomy M., Hilal A. Alternaria rot on plants and fruits of prickly pear in Egypt: Recent outbreak of a destructive disease and its management. *Egyptian Journal of Applied Science 2001*;16:93–107.

Abou-Zaid, F., Ahmed F., Zedan A. E. H. Using of Prickly Pear (Opuntia Spp.) Fruit Juice and Peels in Cookies Production. *Alexandria Science Exchange Journal* 2022;43. https://doi.org/10.21608/asejaiqjsae.2022.239236.

Ajila, C. M., Leelavathi K., Prasada Rao U. J. S. Improvement of dietary fiber content and antioxidant properties in soft dough biscuits with the incorporation of mango peel powder. *J Cereal Sci* 2008;48. https://doi.org/10.1016/j.jcs.2007.10.001.

Albano, C., Negro C., Tommasi N., Gerardi C., Mita G., Miceli A., et al., Betalains, Phenols and Antioxidant Capacity in Cactus Pear [Opuntia ficus-indica (L.) Mill.] Fruits from Apulia (South Italy) Genotypes. *Antioxidants* 2015;4:269–80. https://doi.org/10.3390/antiox4020269.

Alivernini, A., Fares S., Ferrara C., Chianucci F. An objective image analysis method for estimation of canopy attributes from digital cover photography. *Trees - Structure and Function* 2018;32. https://doi.org/10.1007/s00468-018-1666-3.

Allen, R. G., Pereira L. S., Raes D., Smith M. Crop Evapotranspiration: Guidelines for computing crop water requirements. *FAO Irrigation and Drainage Paper* No. 56. Rome: 1998.

Ammar, M., Shltout A., Kamhawy M. Cladode and fruit rots of prickly pear (Opuntia ficus-indica L. Mill.) in Egypt. *Egypt J Phytopathol* 2004;32:119–28.

Attia, E. S. A., Shehata H. A., Askar A. An alternative formula for the sweetening of reduced-calorie cakes. *Food Chem* 1993;48. https://doi.org/10.1016/0308-8146(93)90052-H.

Bakr, A. A. Impact of partial replacement of alfalfa hay with prickly pear peels (PPPS) on productive performance of rabbits. *Egyptian Journal of Nutrition and Feeds* 2019;22:535–50. https://doi.org/10.21608/ejnf.2019.79445.

Barbera, G., Carimi F., Inglese P. Past and present role of the Indian-fig prickly-pear (Opuntia ficus-indica (L.) Miller, Cactaceae) in the agriculture of sicily. *Econ Bot* 1992;46:10–20. https://doi.org/10.1007/bf02985249.

Bendhifi Zarroug, M., Baraket G., Zourgui L., Souid S., Salhi Hannachi A. Genetic diversity and phylogenetic relationship among tunisian cactus species (opuntia) as revealed by random amplified microsatellite polymorphism markers. *Genetics and Molecular Research* 2015;14. https://doi.org/10.4238/2015.February.13.21.

Betancourt, C., Cejudo-Bastante M. J., Heredia F. J., Hurtado N. Pigment composition and antioxidant capacity of betacyanins and betaxanthins fractions of Opuntia dillenii (Ker

Gawl) Haw cactus fruit. *Food Research International* 2017;101:173–9. https://doi.org/10.1016/j.foodres.2017.09.007.

Bousbia, N., Mazari A., Lamoudi L., Akretche-kelfat S., Chibane N., Dif M. E. Evaluation of the Phytochemical Composition and the Antioxidant Activity of Cactus Pear Flowers and Fruit Derivatives. *Revue Agrobiologia* 2022;12:3235–43.

Brian, K. J., Gould J. Michael, Warner K. High-fiber, noncaloric flour substitute for baked foods. Properties of alkaline peroxide-treated lignocellulose. *Cereal Chem* 1989;66:209–13.

CAAES. Cultivation and production of cactus pear. Report no.993/2005. Cairo: Central Administration for Agricultural Extension Services, *Ministry of Agriculture and Land Reclamation;* 2012.

Cassano, A., Conidi C., Drioli E. Physico-chemical parameters of cactus pear (Opuntia ficus-indica) juice clarified by microfiltration and ultrafiltration processes. *Desalination* 2010;250. https://doi.org/10.1016/j.desal.2009.09.117.

Chaiya, B., Pongsawatmanit R. Quality of batter and sponge cake prepared from wheat-tapioca flour blends. *Kasetsart Journal - Natural Science* 2011;45:305–13.

Chavez-Santoscoy, R. A., Gutierrez-Uribe J. A., Serna-Saldívar S. O. Phenolic composition, antioxidant capacity and in vitro cancer cell cytotoxicity of nine prickly pear (Opuntia spp.) juices. *Plant Foods for Human Nutrition* 2009;64. https://doi.org/10.1007/s11130-009-0117-0.

Consoli, S., Inglese G., Inglese P. Determination of Evapotranspiration and Annual Biomass Productivity of a Cactus Pear [Opuntia ficus-indica L. (Mill.)] Orchard in a Semiarid Environment. *Journal of Irrigation and Drainage Engineering* 2013;139. https://doi.org/10.1061/(asce)ir.1943-4774.0000589.

Coria Cayupán, Y. S., Ochoa M. J., Nazareno M. A. Health-promoting substances and antioxidant properties of Opuntia sp. fruits. Changes in bioactive-compound contents during ripening process. *Food Chem* 2011;126. https://doi.org/10.1016/j.foodchem.2010.11.033.

da Silva K. B., de Oliveira J. S., Santos E. M., Ramos J. P. de F., Cartaxo F. Q., Givisiez P. E. N., et al., Cactus pear as roughage source feeding confined lambs: Performance, carcass characteristics, and economic analysis. *Agronomy* 2021;11. https://doi.org/10.3390/agronomy11040625.

De Souza, A. J. A., Dos Santos B. D., Dos Reis A. G., Da Silva A. J., Camelo C. T., Luíz S., et al., Yield and survival rate of "Gigante" cactus pear under regulated deficit irrigation using wastewater. *International Journal of Development Research* 2021;11:49469–76. https://doi.org/10.37118/ijdr.22590.08.2021.

Djeridane, A., Yousfi M., Nadjemi B., Boutassouna D., Stocker P., Vidal N. Antioxidant activity of some algerian medicinal plants extracts containing phenolic compounds. *Food Chem* 2006;97:654–60. https://doi.org/10.1016/J.FOODCHEM.2005.04.028.

Du Toit, A., De Wit M., Naude S., Taljaard M., Fouche H. J., Hugo A, et al., Functional properties and sensory evaluation of mucilage from South-African cactus pear cladodes. *Acta Hortic* 2019;1247. https://doi.org/10.17660/ACTAHORTIC.2019.1247.34.

Duffková, R., Zajíček A., Nováková E. Actual evapotranspiration from partially tile-drained fields as influenced by soil properties, terrain and crop. *Soil and Water Research* 2011;6. https://doi.org/10.17221/20/2010-swr.

Dürüst, N., Sümengen D., Dürüst Y. Ascorbic Acid and Element Contents of Foods of Trabzon (Turkey). *J Agric Food Chem* 1997;45. https://doi.org/10.1021/jf9606159.

Elbana, M., El-Gamal E., Mohamed A., Fernando A. L., Pari L., Outzourhit A., et al., Effect of irrigation scheduling on canopy cover development and crop-water management related parameters of O. ficus-indica under prolonged drought conditions. *Scientific Journal of Agricultural Sciences* 2020a;2. https://doi.org/10.21608/sjas.2020.50551.1061.

Elbana, M., Elwakeel M., Rashad M. On-Farm Water Management and its Impacts on Productivity and Quality of Cactus Pear (opuntia ficus-indica). *Alexandria Science Exchange Journal* 2020b;41. https://doi.org/10.21608/asejaiqjsae.2020.132557.

Elwakeel, M., Hosny M., Rashad M., Elbana M. Utilization of cactus pear fruits' pulp planted under severe drought conditions as source of natural pigments and antioxidant in cake making. *Scientific Journal of Agricultural Sciences* 2021;3. https://doi.org/10.21608/sjas.2021.92321.1148.

FAO and ICARDA. Crop Ecology, *Cultivation and Uses of Cactus Pear*. vol. 20. 2017.

Field, W. P., Collier F. W., Wallingford H. R. *Checklist to assist preparation of small-scale irrigation projects in Sub-Saharan Africa*. 1998.

Filgueiras, R., Almeida T. S., Mantovani E. C., Dias S. H. B., Fernandes-Filho E. I., da Cunha F. F., et al., Soil water content and actual evapotranspiration predictions using regression algorithms and remote sensing data. *Agric Water Manag* 2020;241. https://doi.org/10.1016/j.agwat.2020.106346.

Gomez-Conde, M. S., Garcia J., Chamorro S., Eiras P., Rebollar P. G., Pérez De Rozas A., et al., Neutral detergent-soluble fiber improves gut barrier function in twenty-five-day-old weaned rabbits. *J Anim Sci* 2007;85. https://doi.org/10.2527/jas.2006-777.

Graf, A., Bogena H. R., Drüe C., Hardelauf H., Pütz T., Heinemann G., et al., Spatiotemporal relations between water budget components and soil water content in a forested tributary catchment. *Water Resour Res* 2014;50. https://doi.org/10.1002/2013WR014516.

Hassan, S., Liguori G., Inglese P., Louhaichi M., Sortino G. The effect of soil volume availability on Opuntia ficus-indica canopy and root growth. *Agronomy* 2020;10. https://doi.org/10.3390/agronomy10050635.

Hossain, A., Da Silva J. A. T. Wheat production in Bangladesh: Its future in the light of global warming. *AoB Plants* 2013;5:1–24. https://doi.org/10.1093/aobpla/pls042.

Israelsen, O. W., Hansen V. E. Flow of water into and through soils: Irrigation principles and practices. *Third.* New York: John Wiley and sons, Inc.; 1979.

Jacob, J. K., Hakimuddin F., Paliyath G., Fisher H. Antioxidant and antiproliferative activity of polyphenols in novel high-polyphenol grape lines. *Food Research International* 2008;41. https://doi.org/10.1016/j.foodres.2008.02.009.

Jung, M., Reichstein M., Ciais P., Seneviratne S. I., Sheffield J., Goulden M. L., et al., Recent decline in the global land evapotranspiration trend due to limited moisture supply. *Nature* 2010;467. https://doi.org/10.1038/nature09396.

Karababa, E., Coşkuner Y., Aksay S. Some physical fruit properties of cactus pear (Opuntia spp) that grow wild in the eastern Mediterranean region of Turkey. *Journal of the Professional Association for Cactus Development* 2004;6.

Khalafalla, A. A., Said W. M., Mohamed S. Y., Al Saidi F. Diversity and Vegetation Analysis of Plant species Associated with Prickly Pear (Opuntia ficus indica L. Mill.) in Nile Delta, Egypt. *Journal of Scientific Research in Science* 2016;33:1–21. https://doi.org/10.21608/jsrs.2016.15620.

Kiesling, R. Origen, Domesticación y Distribución de Opuntia ficus-indica. *Journal of the Professional Association of Cactus Development* 1998;3:50–60.

Kobylewski, S., Jacobson M. F. Toxicology of food dyes. *Int J Occup Environ Health* 2012;18. https://doi.org/10.1179/1077352512Z.00000000034.

Korhonen, L., Heikkinen J. Automated analysis of in situ canopy images for the estimation of forest canopy cover. *Forest Science* 2009;55. https://doi.org/10.1093/forestscience/55.4.323.

Koster, R. D., Dirmeyer P. A., Guo Z., Bonan G., Chan E., Cox P., et al., Regions of strong coupling between soil moisture and precipitation. *Science* (1979) 2004;305. https://doi.org/10.1126/science.1100217.

Kumar, K., Singh D., Singh R. S. Cactus Pear : Cultivation and uses. *CIAH/Tech/Pub* No73, Pp 38 ICA R-Central Institute for Arid Horticulture, Bikaner, Rajasthan, India 2018.

Kumar, K., Singh D., Singh R. S. Cactus Pear: Cultivation and uses. *CIAH/Tech./Pub.* No73. Rajasthan: 2018.

Lee, K. J., Lee B. W. Estimating canopy cover from color digital camera image of rice field. *J Crop Sci Biotechnol* 2011;14. https://doi.org/10.1007/s12892-011-0029-z.

Liguori, G., Inglese P. Cactus pear fruit production: orchard planting and management of Opuntia ficus-indica. *Proceedings of International Cactus Pear Workshop 2015*.

Livrea, M. A., Tesoriere L. Health benefits and bioactive components of the fruits from Opuntia ficus-indica [L.] Mill. *Journal of the Professional Association for Cactus Development* 2006;8.

Makwana, A., Sabhadiya H. Determination of Engineering Properties of Cactus Pear (Opuntia). *International Journal of Innovative Research in Engineering & Management (IJIREM)* 2022;9:311–5.

Mashope, B. K. Characterization of cactus pear germplasm in South Africa. *Faculty of Natural and Agricultural Sciences, University of Free State*, 2007.

Michael, A. M. *Irrigation: Theory and Practice. Second.* New Delhi: VIKAS Publishing House PVT LTD; 2006.

Molden, D., Murray-Rust H., Sakthivadivel R., Makin I. A water-productivity framework for understanding and action. *Water productivity in agriculture: limits and opportunities for improvement, 2003*. https://doi.org/10.1079/9780851996691.0001.

Najjar, D. Women's contributions to climate change adaptation in Egypt's Mubarak Resettlement Scheme through cactus cultivation and adjusted irrigation. *A Political Ecology of Women, Water, and Global Environmental Change,* 2018. https://doi.org/10.4324/9781315796208-9.

Negm, A. M. Conventional Water Resources and Agriculture in Egypt. *The handbook of Environmental Chemistry.* vol. 74. Switzerland AG: Springer Nature; 2019. https://doi.org/https://doi.org/10.1007/978-3-319-95065-5.

North, G. B., Nobel P. S. Drought-induced changes in hydraulic conductivity and structure in roots of Ferocactus acanthodes and Opuntia ficus-indica. *New Phytologist* 1992;120. https://doi.org/10.1111/j.1469-8137.1992.tb01053.x.

Osuna-Martínez, U., Reyes-Esparza J., Rodríguez-Fragoso L. Cactus (Opuntia ficus-indica): A Review on its Antioxidants Properties and Potential Pharmacological Use in Chronic Diseases. *Nat Prod Chem Res* 2014;2. https://doi.org/10.4172/2329-6836.1000153.

Palmeri, R., Parafati L., Arena E., Grassenio E., Restuccia C., Fallico B. Antioxidant and antimicrobial properties of semi-processed frozen prickly pear juice as affected by cultivar and harvest time. *Foods* 2020;9. https://doi.org/10.3390/foods9020235.

Passioura, J. B. Water Transport in and to Roots. *Annu Rev Plant Physiol Plant Mol Biol* 1988;39:245–65. https://doi.org/10.1146/annurev.pp.39.060188.001333.

Phocaides, A. (Andreas). Handbook on pressurized irrigation techniques. *Food and Agriculture Organization of the United Nations.*; 2007.

Piga, A. Cactus pear: A fruit of nutraceutical and functional importance. *Journal of the Professional Association for Cactus Development* 2004;6.

Prieto-Santiago, V., Cavia M. M., Alonso-Torre S. R., Carrillo C. Relationship between color and betalain content in different thermally treated beetroot products. *J Food Sci Technol* 2020;57. https://doi.org/10.1007/s13197-020-04363-z.

Rahmati, M., Groh J., Graf A., Pütz T., Vanderborght J., Vereecken H. On the impact of increasing drought on the relationship between soil water content and evapotranspiration of a grassland. *Vadose Zone Journal* 2020;19. https://doi.org/10.1002/vzj2.20029.

Rodríguez, L., Lopez Deibys J., Preston T. R., Peters K. New Cocoyam (Xanthosoma sagittifolium) leaves as partial replacement for soya bean meal in sugar cane juice diets for growing pigs. *Livestock Research for Rural Development* 2006;18.

Romjaun, Z. Z., Prakash J. Development and Assessment of Fibre-Enriched Muffins. *Advances in Food Sciences* 2013;35:159–65.

Saenz, C. Processing technologies: An alternative for cactus pear (Opuntia spp.) fruits and cladodes. *J Arid Environ* 2000;46. https://doi.org/10.1006/jare.2000.0676.

Sakuta, M. Diversity in plant red pigments: Anthocyanins and betacyanins. *Plant Biotechnol Rep* 2014;8. https://doi.org/10.1007/s11816-013-0294-z.

Scarano, P., Sciarrillo R., Tartaglia M., Zuzolo D., Guarino C. Circular economy and secondary raw materials from fruits as sustainable source for recovery and reuse. A review. *Trends Food Sci Technol* 2022;122. https://doi.org/10.1016/j.tifs.2022.02.003.

Seneviratne, S. I., Corti T., Davin E. L., Hirschi M., Jaeger E. B., Lehner I., et al., Investigating soil moisture-climate interactions in a changing climate: A review. *Earth Sci Rev* 2010;99. https://doi.org/10.1016/j.earscirev.2010.02.004.

Sigwela, V., De Wit M., Toit A. du, Osthoff G., Hugo A. Bioactive betalain extracts from cactus pear fruit pulp, beetroot tubers, and amaranth leaves. *Molecules* 2021;26. https://doi.org/10.3390/molecules26165012.

Skalicky, M., Kubes J., Shokoofeh H., Tahjib-Ul-Arif Md, Vachova P., Hejnak V. Betacyanins and Betaxanthins in Cultivated Varieties of Beta vulgaris L. Compared to Weed Beets. *Molecules* 2020;25.

Snyman, H. A. Effect of various water application strategies on root development of Opuntia ficus-indica and O. robusta under greenhouse growth conditions. *Journal of the Professional Association for Cactus Development* 2004;6.

Sumaya-Martínez, M. T., Cruz-Jaime S., Madrigal-Santillán E., García-Paredes J. D., Cariño-Cortés R., Cruz-Cansino N., et al., Betalain, acid ascorbic, phenolic contents and antioxidant properties of purple, red, yellow and white cactus pears. *Int J Mol Sci* 2011;12. https://doi.org/10.3390/ijms12106452.

Tsegay, Z. T., Sathyanarayana C. B., Lemma S. M. Optimization of cactus pear fruit fermentation process for wine production. *Foods* 2018;7. https://doi.org/10.3390/foods7080121.

USDA. *Estimating Soil Moisture by Feel and Appearance*. Washington, DC.: 1998.

Xiong, Y., West C. P., Brown C. P., Green P. E. Digital image analysis of old world bluestem cover to estimate canopy development. *Agron J* 2019;111. https://doi.org/10.2134/agronj2018.08.0502.

Zenteno-Ramírez, G., Juárez-Flores B. I., Aguirre-Rivera J. R., Monreal-Montes M., García J. M., Serratosa M. P., et al., Juices of prickly pear fruits (Opuntia spp.) as functional foods. *Italian Journal of Food Science* 2018;30.

Biographical Sketches

Maha Elbana, PhD

Affiliation:

Department of Soil and Water Sciences, Faculty of Agriculture, Beni-Suef University, Beni Suef, Egypt.

Education:

2005 - B.S. Agricultural Sciences, Alexandria University, Egypt

2007 - Diploma Integrated planning for rural development and environmental management, Instituto Agronómico Mediterráneao de Zaragoza (IAMZ), University of Lleida, Spain

2008 - M.S. Soil and Water Sciences, University of Lleida, Spain

2011 - PhD Soil and Water Sciences, University of Lleida, Spain.

Research and Professional Experience:

Dr. Maha Elbana, Ph.D. is an Associate Professor (Soil Physics & Water Relationships) at Faculty of Agriculture, Beni-Suef University (Egypt) and an

Associate Trainer certified by the International Board of Certified Trainers (IBCT) (Netherlands). Before that, for two years, she served as On-Farm Water Management Specialist at the International Centre for Agricultural Research in the Dry Areas (ICARDA). She holds a Diploma in Rural Development and Environmental Management (2007) from the CIHEAM-Zaragoza (Spain), a M.Sc. (2008) and a Ph.D. (2011) in water resources and irrigation management from the University of Lerida (Spain). She worked as a Research Staff in Training at the University of Girona (Spain) for two years. She is a reviewer for many ISI Journals and has multiple publications in peer reviewed journals dealing with agricultural irrigation and water management. Dr. Maha Elbana research interests are On-farm soil and water management; Treated wastewater application in Agriculture; Soil physics.

Professional Appointments:

2021 - present - Associate Professor, Soil Physics and Water Relationships, Faculty of Agriculture, Beni-Suef University, Egypt
2020-2020 - Visiting Researcher, the Experimental Station of Aula Dei (EEAD), Spanish National Research Center (CSIC), Zaragoza, Spain
2015-2021 - Assistant Professor, Soil and Water Sciences, Faculty of Agriculture, Beni-Suef University, Egypt
2015-2017 - Certified Associate Trainer by the International Board of Certified Trainers (IBCT), the Netherlands
2013-2015 - On-Farm Water Management Specialist, International Center for Agricultural Research in the Dry Areas (ICARDA), Cairo office
2009-2012 - Researcher Staff in Training, University of Girona, Spain.

Honors:

2005 - Abdelmonem Balba Award for outstanding student achievement
2005 - Grant from the Egyptian Ministry of communication to study Office Administration Track in cooperation with Microsoft Office Specialist – Egypt
2006 - Grant from the Mediterranean Agronomic Institute of Zaragoza, International Centre for Advance Mediterranean Agronomic Studies (CIHEAM-IAMZ), Zaragoza-Spain to conduct Diploma of Specialized Post Graduate Studies (DSPU) in Integrated Planning for Rural Development and Agricultural Management

2008 - FPI (Formación del Personal Investigador) Grant from the Spanish Ministry of Science and Innovation to obtain Master and PhD degrees from University of Lleida, Spain.

Publications from the Last 3 Years:

[1] Gamal, R., Abou-Hadid A.F., Omar M.E.D., Elbana M. 2024. "Does climate change affect wheat productivity and water demand in arid regions? Case study of Egypt. *Journal of Agriculture and Food Research,* 16 (101181). https://doi.org/10.1016/j.jafr.2024.101181.

[2] Elwakeel, M., Hosny M., Rashad M., Elbana M. 2021. "Utilization of cactus pear fruits' pulp planted under severe drought conditions as source of natural pigments and antioxidant in cake making." *Scientific Journal of Agricultural Sciences*, 3(2): 264-276. doi: 10.21608/sjas.2021.92321.1148.

[3] Rashad, M., Kenawy E., Hosny A., Hafez M., Elbana M. 2021. "An environmental friendly super absorbent composite based on rice husk as soil amendment to improve plant growth and water productivity under deficit irrigation conditions." *Journal of Plant Nutrition*, 44(7), pp. 1010–1022. https://doi.org/10.1080/01904167.2020.1849293.

Mahmoud Abd El Hameed Elwakeel

Affiliation:

Associate Professor & Head of Food Science Department - Faculty of Agriculture - Beni-Suef University.

Education:

- Ph.D., 2011. Food Science Department, Faculty of Agricultural, Ain Shames University
- M.Sc., 2007. Food Science Department, Faculty of Agricultural, Ain Shames University.
- B.Sc., 2002. Food science department, faculty of Agricultural, Ain Shames University.

Research and Professional Experience:

Food science and technology, Food safety, Food processing.

Professional Appointments:

- Jun. 2022 – current: Associate Professor in Food Science Department - Faculty of Agriculture - Beni-Suef University
- Sept. 2015 – Current: Associate Trainer. Certified by the International Board for Certified Trainers (IBCT), The Netherlands.
- Sept. 2017- June, 2023, Head of Beni-Suef branch. National Food Safety Authority (NFSA), EGYPT
- Oct. 2017 – Jul. 2018, Technical Coordinator. Improving Household Food and Nutrition Security in Egypt by Targeting Women and the Youth project, Food and Agricultural Organization of the United Nations - FAO (Cairo office), Egypt.
- Aug. 2015 – Oct. 2017, Consultant. Improving Household Food and Nutrition Security in Egypt by Targeting Women and the Youth project, Food and Agricultural Organization of the United Nations - FAO (Cairo office), Egypt.
- Jan. 2008 – Sept. 2015, Food Safety Expert and Trainer. Egyptian Food Safety Information Center (EFSIC), Food Technology Research Institute - Agricultural Research Center – Ministry of Agricultural and land Reclamation, Egypt.
- Jan. 2007 – Sept. 2015, Researcher Scientist. Department of Special Food and Nutrition, Food Technology Research Institute - Agricultural Research Center – Ministry of Agricultural and land Reclamation, Egypt

Honors:

- Obtained the first prize for the best " Ph.D degree thesis "in the (Unilever -Mashreq) competition to encourage the scientific research 2012.
- Obtained the first prize for the best "Master degree thesis" in the (Unilever -Mashreq) competition to encourage the scientific research 2008.

Publications from the Last 3 Years:

[1] Elbana M., Elwakeel M., Mohamed A.E., Rashad M. 2020. On-farm water management and its impacts on productivity and quality of cactus pear (*Opuntia*

ficus-indica). *Alexandria Science Exchange Journal* 41(4): 413:522. https://dx.doi.org/10.21608/asejaiqjsae.2020.132557.

[2] Elwakeel M.A. 2020. Preparation of functional juice blends containing bioactive healthy compounds and antitumor effects. *African J. Biol. Sci.,* 16 (1): 285-295.

[3] Ibrahim G., Elwakeel M., Hussein A. 2021. Evaluation of quality attributes, antioxidant activity and volatile compounds of two cactus pear juices blended with guava juice. *Egypt. J. Chem.* Vol. 64, No. 5 pp. 2613-2622 doi: 10.21608/EJCHEM.2021.58555.3263.

[4] Elwakeel M.A., Hosny M., Rashad M. and Elbana M.2021. Utilization of cactus pear fruits' pulp planted under severe drought conditions as source of natural pigments and antioxidant in cake making. *Scientific Journal of Agricultural Sciences* 3 (2): 264-276 doi: 10.21608/sjas.2021.92321.1148.

[5] Ibrahim G., Elwakeel M., Hussein A. 2022. Effect of Blending Ratios from Pineapple Juice on Storage, Physicochemical, Antiradical Activity and Volatile Compounds of Two Cactus Pear Juices. *Letters in Applied NanoBioScience Journal Volume* 11, Issue 4, 2022, 4139-4160 https://doi.org/10.33263/LIANBS114.41394160.

[6] Elwakeel M.A. and Ismael S.2022. Evaluation of quality and Thermal Analysis of Gluten Free Crackers making from cassava and sweet potato. *Middle East Journal of Applied Sciences*. Volume: 012 | Issue: 01.

Mohamed Rashad Abdelfattah

Affiliation:

City of Scientific Research and Technological Applications (SRTA-City)

Education:

Ph.D. Environmental Studies, M.Sc. Environmental Studies, B.Sc. Soil and Water Sciences

Research and Professional Experience:

Sustainable agriculture and green economy.

Professional Appointments:

Professor of Environmental Soil Chemistry, Deputy director for Scientific and Research Affairs, SRTA-City.

Publications from the Last 3 Years:

[1] Elsabagh, S. S., Elkhatib, E. A. & Rashad, M. Novel nano-fertilizers derived from drinkingwater industry waste for sustained release of macronutrients: performance, kinetics and sorption mechanisms. *Sci Rep* 14, 5691 (2024). https://doi.org/10.1038/s41598-024-56274-0.

[2] Elsabagh, S. S., Elkhatib, E. A., & Rashad, M. (2024). Eco-friendly nano-enabledfertilizers derived from date industry waste for sustainable and controlled-releaseof P, K and Mg nutrients: sorption mechanisms, controlled-release performanceand kinetics. *Bioresources and Bioprocessing*, 11(1), 3. https://doi.org/10.1186/s40643-023-00716-6.

[3] Hafez, M., Mohamed Rashad, Ahmed M. Abdallah, and Alexander I. Popov (2023) Towards Green Economy and Agricultural Sustainability Using Environmental Friendly By-products. In book: *Agricultural Research Updates,* Publisher: Nova Publisher.

[4] Hafez, M., Amr El-Nile, Mohamed Ibrahem Hassan et al., (2023) Humic Substances and Their Potential to Enhance Soil, Plants, and Animals' Productivity: A New Concept for Sustainable Agriculture. In book: *Agricultural Research Updates.* Publisher: Nova Publisher.

[5] Rashad, Y., Mohamed Hafez, Mohamed Rashad (2023). Diazotrophic Azotobactersalinestris YRNF3: a probable calcite-solubilizing bio-agent for improving thecalcareous soil properties. *Scientific Reports*, 13 (1): Article number 2062. https://doi.org/10.1038/s41598-023-47924-w.

[6] El-Gamal, E. H. Mohamed Rashad, Maher E. Saleh, Sahar Zaki & Marwa Eltarahony (2023). Potential bioremediation of lead and phenol by sunflower seed husk andrice straw-based biochar hybridized with bacterial consortium: a kinetic study. *Scientific Reports,* 13:21901. https://doi.org/10.1038/s41598-023-49036-x.

[7] Scordia, D., Sebastiano Andrea Corinzia, Jaime Coello, Rosa Vilaplana Ventura, Diana Elisa Jiménez-De-Santiago, Berta Singla Just, Omar Castaño-Sánchez, CarmeCasas Arcarons, Marc Tchamitchian, Léa Garreau, Mohamed Emran, Sami Z. Mohamed, Mai Khedr, Mohamed Rashad, Roxanne Suzette Lorilla, Alexandre Parizel, Giuseppe Mancini, Antonella Iurato, Sergio Ponsá, Corrado Dimauro, FabioGresta, Salvatore Luciano Cosentino, Giorgio Testa (2023). Are agroforestrysystems more productive than monocultures in Mediterranean countries? *Ameta-analysis. Agronomy for Sustainable Development,* 43:73 https://doi.org/10.1007/s13593-023-00927-3.

[8] Mai, K., Emran Mohamed, Gispert Maria, Rashad Mohamed (2023). Immobilizationof Cr3+, Cd2+, and Pb2+ added to calcareous soil amended with composted agroindustrial residues. *Scientific Reports* 13, No. 1 Article No. 8197.

[9] Abu-Saied, M. A., Eman A. El Desouky, Mohamed E. Abou Kamer, Mohamed Hafez, Mohamed Rashad (2023). Influence of magnetic field on the physicochemicalproperties of water molecule under growing of cucumber plant in an arid region, *Journal of King Saud University - Science,* https://doi.org/10.1016/j.jksus.2023.102890.

[10] Rodrigues, C., Camila Damásio de Paula, Soufiane Lahbouki, Abdelilah Meddich, Abdelkader Outzourhit, Mohamed Rashad, Luigi Pari, Isabel Coelhoso, Ana LuísaFernando, and Victor G. L. Souza (2023). Opuntia spp.: An Overview of the BioactiveProfile and Food Applications of This Versatile Crop Adapted to AridLands. *Foods* 12, no. 7: 1465.

[11] Rashad, M., Hafez, M., Popov, A.I., Gaber, H. (2022). Towards sustainable agriculture using extracts of natural materials for transferring organic wastes to environmental- friendly ameliorants in Egypt. *International Journal of Environmental Science and Technology*, http://dx.doi.org/10.1007/s13762-022-04438-8.

[12] Hafez, M., Abdallah, A.M., Mohamed, A.E., Rashad, M. (2022). Influence of environmental-friendly bio-organic ameliorants on abiotic stress to sustainable agriculture in arid regions: A long term greenhouse study in northwestern Egypt. *Journal of King Saud University - Science,* 34(6), 102212. http://dx.doi.org/10.1016/j.jksus.2022.102212.

[13] Rashad, M., Hafez, M., Popov, A. I. (2022). Humic substances composition and properties as an environmentally sustainable system: A review and way forward to soil conservation. *Journal of Plant Nutrition,* 45(7):1072-1122. http://dx.doi.org/10.1080/01904167.2021.2005801.

[14] Hafez, M., Popov, A. I., Rashad, M. (2022). The Biological Correction Using Humic Substances, Vermicompost, and Azospirillum as an Optimum Way of Optimizing Plant Production and Enhancing Soil Micronutrients in Arid Regions. *Open Agricultural Journal*, 16(1): Article number e187433152204180. http://dx.doi.org/10.2174/18743315-v16-e2204180.

[15] Hafez, M., Ge, S., Tsivka, K. I., Popov, A. I., Rashad, M. (2022). Enhancing Calcareous and Saline-Sodic Soils Fertility by Increasing Organic Matter Decomposition and Enzyme Activities: An Incubation Study. Communications in Soil Science and Plant Analysis, In *Press.* http://dx.doi.org/10.1080/00103624.2022.2071930.

[16] El-Refaie, K., Mohamed Rashad, Ali Hosny, Samah Shendy, Dina Gad & Khalil M. Saad-Allah (2022) Enhancement of growth and physiological traits under drought stress in Faba bean (Vicia faba L.) using nanocomposite. *Journal of Plant Interactions*, 17:1, 404-418. https://doi.org/10.1080/17429145.2022.2038293.

[17] Rashad, M., Kenawy, E. R., Hosny, A., Hafez, M., Elbana, M. (2021). An environmental friendly superabsorbent composite based on rice husk as soil amendment to improve plant growth and water productivity under deficit irrigation conditions. *Journal of Plant Nutrition,* Vol (44): 7, 1010-1022. https://doi.org/10.1080/01904167.2020.1849293.

[18] Hafez, M., Abo El-Ezz, S. F., Popov, A. I., & Rashad, M. (2021). Organic Amendments Combined with Plant Growth-Promoting Rhizobacteria (*Azospirillum Brasilense*) as an Eco-Friendly By-Product to Remediate and Enhance the Fertility of Saline Sodic- Soils in Egypt. *Communications in Soil Science and Plant Analysis,* 1-18.

[19] Hafez, M., E. A. Mohamed., Rashad, M., Popov, A. I. (2021) The efficiency of application of bacterial and humic preparations to enhance of wheat (Triticum

aestivum L.) plant productivity in the arid regions of Egypt. *Biotechnology Reports,* Vol (29): e00584. https://doi:10.1016/j.btre 2020.e00584.

[20] Hafez, M., Popov, A. I., Rashad, M. (2021) Integrated use of bio-organic fertilizers for enhancing soil fertility–plant nutrition, germination status and initial growth of corn (Zea Mays L.), *Environmental Technology & Innovation,* Vol (21):101329. https://doi.org/10.1016/j.eti.2020.101329.

Chapter 4

Smart Irrigation: The Convergence of AI and Agriculture

Atin Kumar[1,*]
Nitish Karn[1]
Rohit Kumar[2]
Mohamed A. E. AbdelRahman[3]
and Sharad Sachan[4]
[1]School of Agriculture, Uttaranchal University, Dehradun, Uttarakhand, India
[2]Faculty of Agricultural Sciences, GLA University, Mathura, Uttar Pradesh, India
[3]Division of Environmental Studies and Land Use, National Authority for Remote Sensing and Space Sciences (NARSS), Cairo, Egypt
[4]School of Agriculture, Lovely Professional University, Phagwara (Punjab), India

Abstract

This book chapter explores the convergence of artificial intelligence (AI) and agriculture in the context of smart irrigation systems. The global challenge of water scarcity needs sustainable agricultural practices, and AI technologies offer promising solutions to optimize water usage and improve crop yields. Traditional irrigation methods are inefficient and environmentally concerning, leading to wastage and depletion of natural resources. Smart irrigation systems leverage advanced algorithms like Convolutional Neural Networks (CNN) and Recurrent Neural Networks (RNN) to tailor irrigation schedules, enhance water distribution detection, and improve decision-making in agriculture. Economic and environmental benefits are achieved through these systems, promoting

* Corresponding Author's Email: atinchaudhary0019@gmail.com

In: Irrigation Management
Editors: Maha Elbana and Youssef Brouziyne
ISBN: 979-8-89530-318-4

sustainable farming practices. However, challenges such as data privacy, infrastructure limitations, and the need for farmer education must be addressed for widespread adoption. The potential of AI in agriculture through smart irrigation systems is vast, revolutionizing water management practices and fostering a more sustainable future for the industry.

Keywords: smart irrigation, artificial intelligence, agriculture, water management, sustainability, crop productivity, environmental

List of Abbreviations

AI	Artificial Intelligence
AIoT	Artificial intelligence of things
CCPA	California Consumer Privacy Act
CEP	Complex event processing
CNN	Conventional neural networks
DQN	Prediction-based deep Q-learning
ET	Evapotranspiration
ET_c	Crop evapotranspiration
ET_o	Reference evapotranspiration
FUWARE	Future Internet Ware
GDPR	General Data Protection Regulation
IoF2020	European project Internet of Food and Farm 2020
IoT	Internet of Things
ML	Machine learning
MQTT	Message queuing Technology Transport
MR-CNN	Multi-region conventional neural networks
QUHOMA	Qualitative horticulture marketplace
R-CNN	Raiyani's Mask conventional neural networks
RNN	Recurrent neural networks
TLS	Transport layer security
UAV	Unmanned aerial vehicle
VGG	Visual Geometry Group
VPNs	Virtual private networks
VRI	Variable rate irrigation.

Introduction

The Global Challenge: Water Scarcity and the Need for Sustainable Agriculture

The global challenge of water scarcity presents a significant hurdle for the agriculture industry. As water resources become increasingly strained, the need for sustainable agriculture practices is more crucial than ever. The intersection of artificial intelligence and agriculture holds the potential to enhance water efficiency and increase crop yields. Through the integration of AI (Artificial Intelligence), farmers gain the ability to make better-informed choices regarding irrigation, leading to more sustainable and effective agricultural methods. This paper aims to examine the possibilities presented by smart irrigation systems and to investigate how AI innovations can transform the management of water resources in agriculture (Ramachandran et al., 2022).

Traditional Irrigation Practices: Inefficiencies and Environmental Concerns

Traditional irrigation practices have been the cornerstone of agriculture for centuries, but they come with inherent inefficiencies and environmental concerns. One of the main inefficiencies is the imprecise application of water, which leads to wastage and uneven distribution across fields. Additionally, the reliance on manual observation and decision-making often results in overwatering or underwatering, both of which can detrimentally impact crop health and yield.

Moreover, traditional irrigation methods can contribute to environmental concerns such as soil erosion, salinization, and depletion of natural water resources. The excessive use of water in irrigation can lead to the buildup of salts in the soil, affecting its fertility and long-term productivity. Furthermore, water extraction from rivers and aquifers for irrigation purposes can disrupt delicate ecosystems and reduce the water availability for other essential uses (Adli et al., 2023).

As we explore the difficulties associated with conventional irrigation methods, it becomes clear that incorporating AI technologies presents significant opportunities to tackle these challenges and advance toward more sustainable and effective agricultural techniques. By adopting smart irrigation systems driven by AI, farmers can surpass the constraints of traditional approaches and enhance water utilization while reducing environmental harm.

The Promise of Smart Irrigation: AI and Data-Driven Solutions

The potential of smart irrigation lies in combining AI and data-driven approaches to transform water management in agriculture. By utilizing AI technologies and up-to-date data, smart irrigation systems can accurately customize water distribution according to the unique requirements of crops, soil conditions, and weather variations. This precise strategy enhances water usage efficiency and guarantees optimal crop development and productivity (García et al., 2020).

AI algorithms can examine various elements, such as soil moisture levels, weather predictions, and plant physiology, to make educated choices regarding irrigation timing and water distribution. This high level of accuracy not only saves water but also reduces the environmental consequences linked with over-irrigation, such as soil salinity and depletion of water resources (Ali et al., 2023).

Moreover, smart irrigation systems facilitate ongoing monitoring and adjustment, empowering farmers to remotely oversee and enhance water usage. This automation decreases dependency on manual involvement, thereby lowering the risk of human error and improving operational efficiency.

Understanding Smart Irrigation

Components of a Smart Irrigation System

A smart irrigation system typically consists of several key components (Figure 1) that work together to optimize water usage and enhance crop productivity. A smart irrigation system's primary components include:

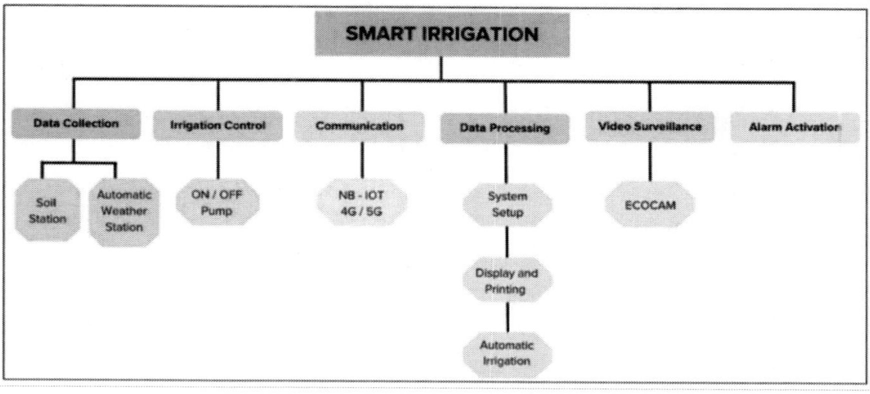

Figure 1. Diagram showing the different components and functions of smart irrigation systems (Ali et al., 2023).

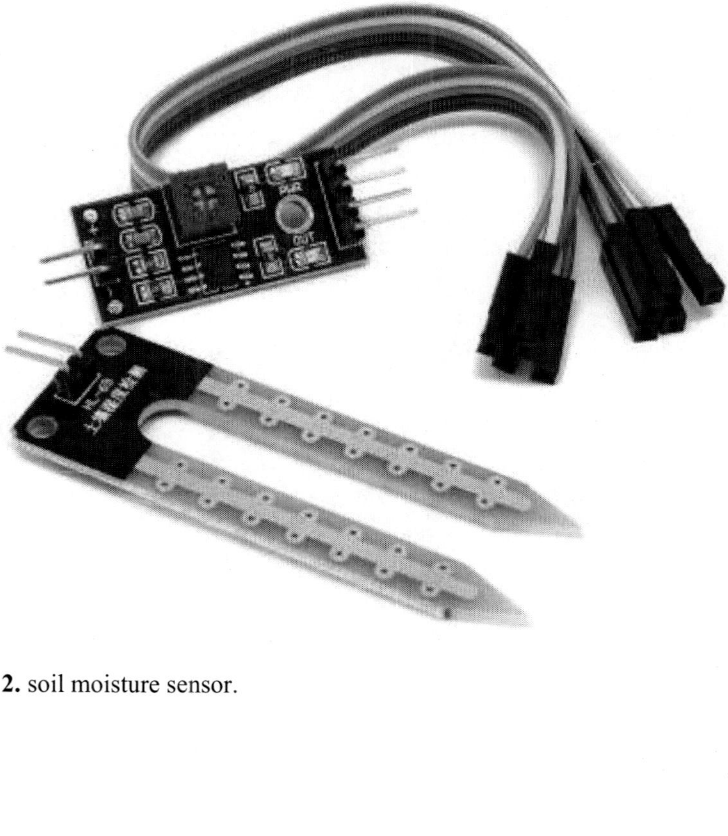

Figure 2. soil moisture sensor.

1. Sensors: Soil moisture sensors (Figure 2), weather sensors, and other environmental sensors are essential components of a smart irrigation system. Soil moisture sensors gauge soil moisture levels, guiding the system in determining the timing and quantity of water application. Weather sensors offer immediate data on temperature, humidity, wind speed, and rainfall, aiding in adjusting irrigation schedules according to present weather conditions (Ali et al., 2023).
2. Controller: The controller serves as the core of the smart irrigation system, interpreting sensor data to establish the irrigation schedule, including timing, duration, and frequency. It operates based on preset parameters and sensor inputs and can be programmed manually or automatically, depending on system settings (Abioye et al., 2022).
3. Actuators: Actuators are devices that physically control water flow in the irrigation system. They can include valves, pumps, and sprinklers that open and close to deliver water to the crops. Actuators are controlled by the irrigation system's controller based on the irrigation schedule and sensor readings (Ramachandran et al., 2022).
4. Communication Network: A communication network is essential for transmitting data between the various components of the smart irrigation system. This network can be wired or wireless and allows sensors, controllers, and actuators to communicate with each other in real time. Communication networks enable remote monitoring and control of the irrigation system, providing farmers access to critical data and system settings from anywhere. (Altalak et al., 2022)
5. User Interface: Farmers can communicate with the smart irrigation system through the user interface. It can be a mobile app, a web-based dashboard, or a control panel that allows users to monitor system performance, adjust settings, and receive alerts or notifications. The user interface gives farmers visibility and control over their irrigation system, enabling them to make informed decisions and optimize water management practices.
6. Data Analytics and Decision Support: Smart irrigation systems incorporate data analytics and decision-support tools to analyze sensor data, weather forecasts, and past patterns. These tools aid farmers in data-informed decisions regarding irrigation scheduling, water management, and crop health. By leveraging these capabilities, farmers can enhance water utilization, minimize waste, and boost crop yields.

7. A smart irrigation system includes sensors, controllers, actuators, a communication network, a user interface, and data analytics tools. These collaborate to observe soil moisture, weather conditions, and crop water needs. The system automates irrigation scheduling and enhances water utilization for sustainable and efficient crop cultivation.

Types of Smart Irrigation Technologies: Drip Irrigation, Sprinklers, Subsurface Irrigation

Smart irrigation technologies aim to improve water efficiency, boost crop yields, and encourage sustainable farming practices. These innovations utilize sensors, controllers, actuators, communication networks, and data analytics tools to monitor soil moisture, weather conditions, and crop water requirements. They automate irrigation scheduling to maximize water usage for sustainable crop production.

Soil Moisture Sensors

Soil moisture sensors are a prevalent form of smart irrigation technology. The sensors gauge soil moisture levels, offering immediate insights into soil conditions. By monitoring soil moisture, farmers ascertain the appropriate timing and quantity of water application, guaranteeing optimal water provision for crops. These sensors can be placed in the soil at various depths to improve the accuracy of measuring soil moisture. (Vallejo-Gómez et al., 2023).

Weather Stations

Weather stations are integral to smart irrigation systems, gathering information on temperature, humidity, wind speed, and rainfall. This data enables farmers to adapt irrigation schedules according to prevailing weather conditions, optimizing water use and reducing waste. Weather stations can be installed onsite or accessed remotely via online weather services (Sami et al., 2022).

Evapotranspiration (ET) Controllers

Evapotranspiration (ET) controllers are a form of smart irrigation technology that utilizes weather data, such as temperature and humidity, to compute the rate of water evaporation from soil and transpiration from plants. By adapting irrigation schedules according to ET rates, farmers can provide sufficient water to crops while conserving resources. When adjusting irrigation schedules, these controllers can be customized to accommodate various crop types, soil compositions, and weather patterns (Montesano et al., 2018).

Drip Irrigation

Drip irrigation systems minimize water loss from evaporation and runoff by using emitters and tubing to feed water directly to plant roots. This targeted method conserves water, making drip irrigation an efficient solution for crop irrigation. Automated controllers adjust water flow and timing based on soil moisture and weather conditions (Kanmani et al., 2021).

Sprinkler System

Sprinkler systems are another common type of smart irrigation technology that uses sensors and controllers to deliver water evenly and efficiently to crops. (Altalak et al., 2022) By adjusting sprinkler direction, flow rate, and timing based on environmental conditions, these systems help optimize water distribution and lower water loss. Sprinkler systems can be automated using controllers that adjust the flow rate and timing of water delivery based on soil moisture and weather conditions.

Remote Monitoring Control

These technologies allow farmers to access up-to-date information on soil moisture, weather, and irrigation schedules via mobile devices or computers. This enables farmers to adapt irrigation settings to changing conditions swiftly, optimizing water management practices. Integration with other smart irrigation technologies, like soil moisture sensors and weather stations, offers a complete overview of irrigation conditions.

Data Analytics and Decision Support Tools

Data analytics and decision-support tools are now commonly incorporated into smart irrigation systems to analyze sensor data, weather predictions, and past patterns. They aid farmers in making informed choices about irrigation scheduling, water management, and crop health. By utilizing these capabilities, farmers can enhance water efficiency, minimize waste, and boost crop yields.

Fuzzy Logic Controllers

Fuzzy logic controllers, a smart irrigation technology, emulate human decision-making using linguistic variables to adjust irrigation settings based on uncertain data, enhancing efficiency. Smart irrigation technologies integrate sensors, controllers, communication networks, and analytics tools to optimize water usage, conserve resources, and boost crop yields sustainably. Adoption of these technologies empowers farmers to enhance irrigation practices, minimize water waste, and promote sustainable agriculture.

Role of AI and Machine Learning: Data Analysis, Predictive Modeling, Optimization

Utilizing artificial intelligence (AI) and machine learning, the Smart Irrigation System leverages the Internet of Things (IoT) to improve data analysis, predictive modelling, and optimization. It collects data from several sensors, including crop, weather, and soil moisture, to monitor crop health and environmental conditions. Utilizing AI algorithms, the system analyzes this data to recognize patterns and trends, aiding in informed decisions regarding irrigation scheduling and resource management (Ramachandran et al., 2022).

The Smart Irrigation System harnesses AI and machine learning for predictive modelling, utilizing historical and current data to forecast weather patterns and crop water needs. This optimizes irrigation schedules and resource distribution, promoting water efficiency and sustainability in agriculture (Albuquerque et al., 2020).

AI and machine learning in the Smart Irrigation System are vital for optimization. The system adjusts irrigation schedules and resource allocation by analyzing real-time sensor data. Machine learning algorithms learn from past patterns to optimize future irrigation schedules based on crop needs and environmental factors, enhancing resource management and crop yield.

Incorporating AI and machine learning enables the Smart Irrigation System to function independently and respond to changing environmental conditions. This fosters sustainable and economical agricultural practices, enhancing productivity while preserving resources.

Data Analysis

Within the IoT-based Smart Irrigation System, data analysis entails scrutinizing extensive data sets obtained from diverse sensors like soil moisture, weather, and crop sensors. The system analyzes this data to uncover patterns, trends, and connections, offering valuable insights into environmental conditions and crop well-being (Badreldeen et al., 2022). The system thoroughly identifies variables like soil moisture, temperature, humidity, and crop water needs by analyzing data. This analysis is based on making well-informed decisions about when to water and how to distribute resources.

Predictive Modeling

Predictive modeling in the Smart Irrigation System involves using historical and real-time data to forecast future environmental conditions and crop water requirements. The system can use machine learning and artificial intelligence to predict weather patterns, soil moisture trends, and crop water needs. These predictive models enable the system to anticipate changes in environmental conditions and make proactive decisions about irrigation scheduling. For example, the system can forecast upcoming dry periods and adjust irrigation schedules accordingly to ensure that crops receive adequate water. Predictive modeling contributes to efficient resource management and helps reduce water waste by aligning irrigation activities with anticipated environmental conditions (Albuquerque et al., 2020).

Optimization

Optimization in the Smart Irrigation System involves refining irrigation schedules and resource distribution in real time using predictive models and machine learning algorithms. These algorithms analyze past irrigation data and environmental factors to dynamically adjust schedules, water flow rates, and resource allocation based on soil moisture, weather forecasts, and crop needs. This adaptive optimization strategy enhances system efficiency, maximizing crop yield while conserving resources (Kanmani et al., 2021).

Smart Irrigation Techniques

Climate change has significantly disrupted the water cycle, leading to more severe droughts and making efficient water use critical in agriculture. The focus on conserving water has heightened, especially in regions prone to annual droughts. Traditional irrigation systems are ineffective in saving water and fail to supply water efficiently (Kanuru et al., 2021). Smart irrigation efficiently addresses water scarcity issues by promoting efficient water use and conservation for future needs. It also reduces input costs for farmers by eliminating the need for manual observation and scheduling. Unlike manual methods, sensor-assisted systems detect soil moisture levels and initiate irrigation, while decision-support-system-assisted systems integrate various sensors to monitor water requirements and regulate irrigation accordingly (Figure 3). Technological advancements have facilitated the evolution of precise and efficient irrigation systems with the advent of decision-support-assisted systems (Valsan et al., 2022). This consists of climate and soil moisture sensors in addition to a cloud-based system for data monitoring in real time. Dynamic simulation models are used to quantify productivity and evaluate the effects of irrigation on crop growth. An optimal irrigation quantity is determined and delivered through an automatic irrigation control system, which can be accessed via a mobile app or remote control. This system is linked to a modern irrigation hub for easy understanding and usage (Muangprathub et al., 2019). The system is based on data analysis and yield predictions. Figure 4 presents a plan for an advanced irrigation hub.

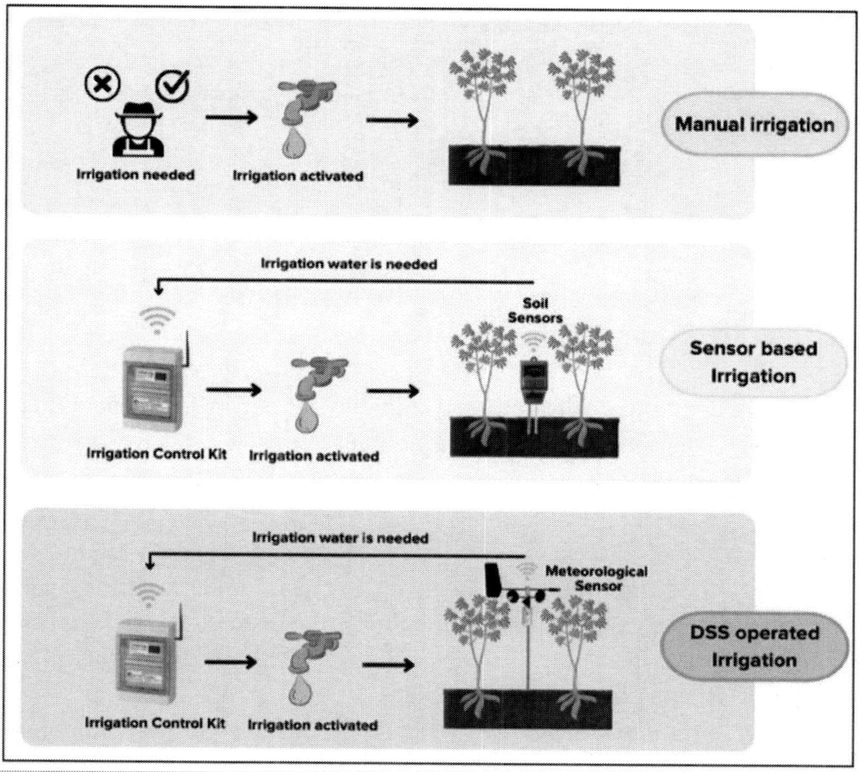

Figure 3. Irrigation applications based on manual versus smart sensor-based and decision support system-assisted (Ali et al., 2023).

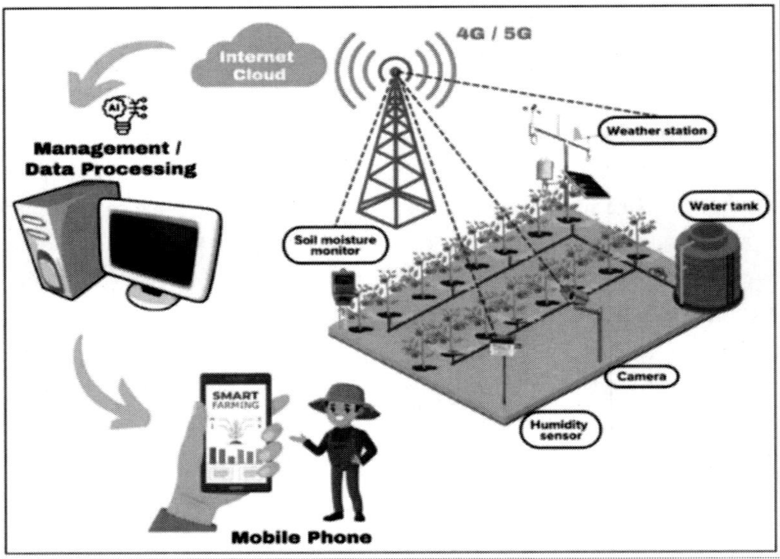

Figure 4. A decision support system, real-time weather and moisture sensors, and an advanced irrigation hub (Ali et al., 2023).

Benefits of Smart Irrigation

Increased Water Efficiency: Reduced Water Waste and Improved Crop Yields

By utilizing a variety of methods and technologies, smart irrigation systems significantly contribute to increasing agricultural yields, decreasing water waste, and boosting water efficiency:

Precision Water Delivery: To precisely determine irrigation timing and quantity and prevent both overwatering and underwatering, smart irrigation systems use sensors to monitor soil moisture, weather, and crop water needs in real time. This maximizes crop yields while minimizing water wastage. (Altalak et al., 2022).

Automated Irrigation Scheduling: Smart irrigation systems automate irrigation based on predefined parameters and data inputs, removing the necessity for manual intervention and guesswork. This saves time and labor and applies to water efficiently, fostering robust plant growth and maximizing yield potential (Sami et al., 2022).

Water Conservation Features: Water-saving features like micro-sprinklers, soil moisture sensors, and drip irrigation are often integrated into smart irrigation systems. Drip irrigation reduces runoff and evaporation by directing water distribution to plant roots, while micro-sprinklers ensure uniform water distribution, reducing losses from wind drift or overspray. Soil moisture sensors optimize irrigation timing and duration, preventing waterlogging and ensuring plants receive ideal moisture levels for growth.

Data-Driven Decision Making: Smart irrigation systems use predictive modelling and data analytics to make well-informed irrigation decisions. These systems dynamically modify irrigation schedules to maximize water usage efficiency, preserve resources, and increase agricultural yields by evaluating historical data, weather forecasts, and crop needs. (Montesano et al., 2018).

Remote Monitoring and Control: Numerous smart irrigation systems provide remote monitoring and control, enabling farmers to access real-time data and adjust settings remotely via mobile devices or computers. This connectivity enables swift responses to changing conditions, troubleshooting, and refining irrigation strategies for optimal outcomes, ensuring efficient water management and sustainable crop production (Badreldeen et al., 2022).

In summary, smart irrigation systems significantly enhance water efficiency, minimize waste, and boost crop yields through precise water delivery, automated scheduling, conservation features, data-driven decision-making, and remote monitoring. Leveraging technology and data analytics, farmers can optimize irrigation practices, conserve water, and improve the sustainability and productivity of their farms.

Improved Crop Health: Precise Water Delivery Based on Real-Time Needs

Smart irrigation systems use modern technology like sensors, data processing, and automation to distribute water to crops according to their actual needs. This tailored approach enhances crop health significantly. They are:

1. Real-time Monitoring: In real-time, smart irrigation systems constantly observe essential factors like soil moisture, weather patterns, and plant hydration needs. This ongoing surveillance

enables the system to adapt irrigation timing as necessary, aligning with the immediate requirements of the crops.

2. Optimized Water Usage: By examining live data, smart irrigation systems can accurately gauge the precise water requirements of crops at any given moment. This precision in delivering water helps mitigate the common problems of both excessive and inadequate watering encountered in traditional irrigation approaches.
3. Prevention of Water Stress: Smart irrigation systems prevent waterlogging ,root diseases caused by overwatering, drought stress, and decreased crop yields due to underwatering. They achieve this by providing crops with optimal water amounts when necessary, thereby reducing water stress and fostering healthier plant development.
4. Maximized Crop Yields: Smart irrigation systems maximize crop yields by delivering the ideal amount of water when needed, which is crucial for various stages of plant growth, flowering, and fruit development, ultimately leading to increased agricultural productivity.
5. Resource Efficiency: Smart irrigation systems enhance crop health and resource efficiency by delivering water precisely and promoting sustainable water management in agriculture through reduced wastage and optimal hydration for plants. This results in healthier growth and improved yields.

Enhanced Decision-Making: Data-Driven Insights for Optimal Irrigation Practices

Smart irrigation systems provide enhanced decision-making through data-driven insights for optimal irrigation practices. By leveraging data analytics and real-time monitoring, these systems offer several benefits for agriculture:

1. Data-Driven Insights: Smart irrigation systems gather and analyze data from diverse sources, such as soil moisture sensors, weather predictions, and crop water demands. This provides valuable insights into crop irrigation requirements and empowers farmers to make well-informed choices.
2. Optimal Irrigation Practices: Farmers can harness data-driven insights to identify the most suitable irrigation tactics for their particular crops and environmental circumstances. These insights

facilitate the refinement of irrigation schedules, water application rates, and duration to meet the plants' requirements precisely.

3. Precision Agriculture: Smart irrigation systems facilitate precision agriculture by furnishing farmers with comprehensive data concerning soil conditions, plant well-being, and water accessibility. This enables personalized irrigation management based on the unique requirements of different crops or field segments.
4. Resource Efficiency: Farmers may increase water efficiency by minimizing water wastage and limiting overwatering by utilizing data-driven insights. By distributing precise amounts of water at ideal times, smart irrigation technologies help agriculture practice sustainable water management.
5. Improved Crop Performance: Utilizing data-driven insights, smart irrigation systems enhance crop performance, yield, and produce quality by optimizing irrigation practices. They enable farmers to make informed decisions, resulting in efficient water utilization, improved crop productivity, and sustainable agriculture.

Environmental Benefits: Reduced Water Footprint and Carbon Emissions

Smart irrigation systems offer environmental benefits by reducing the water footprint and carbon emissions associated with traditional irrigation practices. Here's how these systems contribute to environmental sustainability:

1. Water Conservation: Smart irrigation systems conserve water resources by providing exact water amounts according to real-time data and plant requirements, thus preventing overwatering and reducing water waste. This aids in minimizing the water footprint of agricultural activities through optimized water usage.
2. Efficient Water Management: Smart irrigation systems save water and enhance water efficiency by delivering precise water amounts to crops and optimizing irrigation practices. This decreases water usage per crop yield unit, improving overall water-use efficiency.
3. Energy Savings: Smart irrigation systems integrate energy-efficient elements like low-energy pumps and sensors, diminishing the energy usage linked with irrigation. Through optimizing irrigation schedules

and water delivery, these systems assist farmers in conserving energy and reducing their environmental impact.

4. Reduced Chemical Runoff: Smart irrigation systems support focused irrigation methods, aiding in minimizing the runoff of fertilizers and pesticides into waterways. This enhances water quality and lessens the ecological effects of farming on nearby ecosystems.
5. Climate Resilience: Smart irrigation systems can bolster agricultural systems' resilience to climate change by efficiently managing water and minimizing water consumption. Farmers can mitigate the effects of climate variability on crop production by adjusting irrigation practices in response to changing weather conditions and water availability.

Smart irrigation systems are essential for advancing environmental sustainability in agriculture, as they decrease water usage, enhance water efficiency, conserve energy, reduce chemical runoff, and bolster resilience to climate change. Their adoption fosters a sustainable and eco-friendly approach to irrigation, benefiting agricultural productivity and the environment.

Figure 5. Sensors and Components of the QUHOMA platform (Ali et al., 2023).

Case Studies and Applications

Examples of Successful Smart Irrigation Implementations around the World

Case Study 1

To achieve precision farming, the European project Internet of Food and Farm 2020 (IoF2020) looks into how Internet of Things (IoT) technology may be applied in Europe's food and agricultural industry. Through IoT-driven precision agriculture, farmers can enhance operational efficiency, boost crop yields, and reduce expenses related to inputs like water, fertilizers, insecticides, and herbicides. The QUHOMA (Qualitative Horticulture Marketplace) platform, a smart irrigation solution supported by FIWARE (Future Internet Ware), has been developed and implemented as part of this project (Figure 5). The irrigation schedule is determined using a sophisticated AI approach called complex event processing (CEP), which relies on current irrigation algorithms and daily sensor data accumulation. However, interruptions or inaccuracies in sensor data due to malfunctions or loss of connection can lead to deviations from the optimized target. To address this matter, AI-based predictive analytics were employed, integrating collected data with other public sources to train models tailored to the field. This enables two scenarios to be managed: prolonged unavailability of sensor data using historical data and predefined irrigation schedules for calculating and detecting sensor data anomalies through predictive analytics to facilitate proactive maintenance.

Case Study 2

In 2020, real-world irrigation trials commenced in Cyprus and Slovenia. For instance, one experiment involved cultivating strawberries for approximately 100 days in a tunnel farm with Clay Loam soil in the coastal Ammochostos district of Cyprus (Figure 6). To do this experiment, the QUHOMA platform's smart irrigation scheduling algorithm was modified to include key climate zone values, such as crop evapotranspiration (ET_c), reference evapotranspiration (ET_o), and other important irrigation characteristics. Numerous tasks were finished, such as cleaning up meteorological data, assessing its quality, and determining how inputs affect water retention. For the experiments, algorithms and models underwent calibration and adjustment. When farmers used the QUHOMA irrigation platform instead of the empirical irrigation scheduling tool, water consumption decreased by 10.88% (Moysiadis et al., 2021).

Case Study 3

Kanmani et al., (2021) develops an intelligent irrigation system that leverages CNN (Convolutional Neural Network) algorithms for image processing and data analysis to optimize water usage, increase crop yield, and minimize manual intervention in agricultural operations. The system integrates sensors, water system pumps, and MATLAB code for soil crack identification and irrigation rate determination. CNN models like CaffeNet, LeNet, and AlexNet are employed for plant detection, disease identification, and pattern recognition. Integrating CNN and RNN (Recurrent Neural Network) algorithms into the intelligent system yields notable fruit and vegetable cultivation enhancements, fostering more efficient water usage practices and soil preservation methods. This approach leads to heightened production and enhances agricultural development. The research advocates for smart farming via the Modern Irrigation System, which employs CNN alongside IoT sensors and machine learning (ML) algorithms. This integration enhances water management, increases crop yield, and provides timely weather forecasts, effectively tackling significant agricultural hurdles. The research underscores the transformative potential of CNN and IoT technologies in revolutionizing traditional farming. The Modern Irrigation System, optimizing water usage and providing real-time data analysis, shows promise for sustainable and efficient agriculture. Further research, field trials, and collaboration with farmers are recommended to validate scalability and enhance its impact on crop production and water conservation.

Case Study 4

Prati et al., (2020) investigates the application of innovative technologies to optimize irrigation practices in agriculture. The study involves collecting data using a customized UAV (Unmanned Aerial Vehicle) quadrotor and a Nikon D330 camera, annotating the images using the VGG (Visual Geometry Group) Image Annotator tool, and training a neural network using an advanced version of Raiyani's Mask R-CNN algorithm. The model demonstrates improved accuracy in detecting water jets and curtains in various irrigation systems, showcasing the network's generalization and inference quality. The research underscores the potential uses of the training approach for smart irrigation systems, underscoring the significance of technological progress in enhancing water management efficiency in agriculture.

Highlighting the Impact on Different Crops and Farming Systems

Smart irrigation systems have proven their effectiveness in various agricultural settings. For example, in practical irrigation trials conducted in Cyprus and Slovenia, these systems have been instrumental in optimizing water usage for specific crops like strawberries. Using platforms such as QUHOMA, significant reductions in water usage, such as a 10.88% decrease compared to empirical irrigation schedules, have been achieved. This shows how intelligent irrigation systems can customize watering schedules to meet the unique requirements of individual crops, resulting in more effective water use.

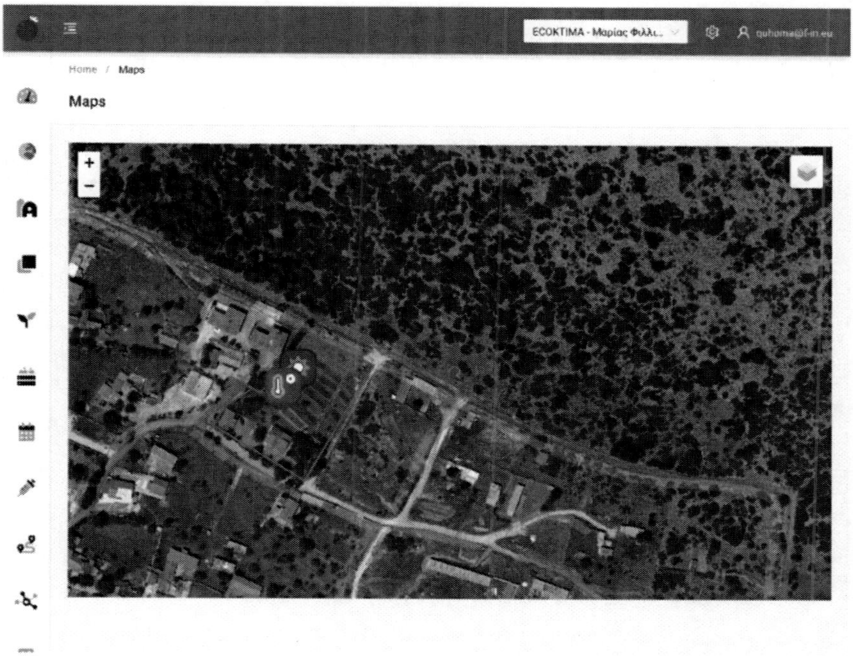

Figure 6. QUHOMA platform implementation on a tunnel farm (Ali et al., 2023).

Moreover, it has been shown that combining innovative algorithms like recurrent neural networks (RNN) and convolutional neural networks (CNN) can improve crop yield and water management. These systems can improve pattern recognition, disease identification, and plant detection using CNN algorithms for image processing and data analysis. This improves fruit and vegetable cultivation and minimizes manual intervention in agricultural operations. RNN algorithms further enhance these systems' predictive capabilities, enabling better decision-making in irrigation practices.

Moreover, applying technologies, such as UAV quadrotors and neural networks, has revolutionized irrigation practices by improving the accuracy of detecting water distribution in various irrigation systems. By training neural networks with annotated images captured by UAVs, these systems can identify water jets and curtains with enhanced precision. This technological advancement underscores the potential of smart irrigation systems in optimizing water management across different farming systems, resulting in more productive and sustainable farming methods.

In summary, smart irrigation systems use advanced technologies and algorithms to maximize water use, increase crop productivity, and minimize manual intervention in agricultural operations. By adjusting irrigation schedules to meet the specific requirements of various crops, enhancing data analysis through CNN and RNN algorithms, and improving the accuracy of water distribution detection using UAVs and neural networks, these systems offer a comprehensive solution for addressing key agricultural challenges and promoting sustainable farming practices.

Challenges and Future Directions

Data Privacy and Security Concerns in Smart Agriculture

This requires a more profound exploration to understand the complexities and implications involved. Here is a more in-depth analysis of data privacy and security concerns in smart agriculture:

Data Privacy Concerns

Personal Data Protection

Smart agricultural systems gather diverse data, including personal details about farmers like their names, contact information, and land ownership. Safeguarding this personal data against unauthorized access or misuse is essential to adhere to privacy regulations and uphold trust among stakeholders.

Data Ownership and Control

Defining ownership rights to data is crucial in smart agriculture. Farmers should possess authority over the data produced by IoT devices on their farms and understand its usage by service providers or other entities. Formulating explicit agreements on data ownership can mitigate conflicts and promote transparency.

Data Sharing and Consent

Maintaining data confidentiality and obtaining consent are paramount concerns when sharing agricultural data with third parties like Agri-tech firms or research institutions. Farmers should be fully informed about the intentions and purposes of data sharing and retain the right to grant or withhold consent accordingly.

Security Challenges

Cybersecurity Risks

IoT devices used in smart agriculture face susceptibility to cyber threats such as malware, ransomware, and unauthorized entry. For instance, if an irrigation system is compromised, it could result in either excessive or insufficient watering of crops, affecting their health and productivity. Enacting robust cyber security protocols, including encryption, authentication, and intrusion detection systems, is imperative to alleviate these vulnerabilities.

Data Integrity and Authenticity:

Securing the accuracy and genuineness of data obtained from IoT devices is essential for informed decision-making in agriculture. Any manipulation of sensor data or irrigation schedules could lead to inadequate crop management practices. Employing validation processes for data and adopting secure storage practices are vital to preserving data integrity.

Network Security

Securing the communication networks used by IoT devices is crucial to thwart unauthorized access and data interception. Safe communication protocols such as Virtual Private Networks (VPNs) and Transport Layer Security (TLS) can help protect data in transit and prevent network-related intrusions.

Compliance and Regulations

Data Protection Regulations
Complying with data protection laws like GDPR (General Data Protection Regulation) in Europe or CCPA (California Consumer Privacy Act) in the United States is crucial for responsible agricultural data management. Adherence to these regulations entails enforcing data protection protocols, acquiring consent for data processing, and upholding the rights of data subjects (Kumawat et al., 2022).

Industry Standards and Best Practices
Adhering to industry norms and recommended procedures for data privacy and security in smart agriculture can help organizations reduce risks and highlight their dedication to safeguarding sensitive data. Standards like ISO/IEC 27001 for managing information security and ISO/IEC 27002 for cybersecurity recommendations offer frameworks for establishing strong security measures (Kumar et al., 2022).

Risk Management

Risk Assessment and Mitigation

Regularly performing risk assessments to pinpoint potential weaknesses in smart agriculture systems and applying measures to mitigate these risks can assist organizations in preemptively tackling security threats. This involves recognizing security deficiencies, ranking risks, and deploying controls to minimize the chances of incidents (Kumar et al., 2017).

Incident Response Planning

Creating a tailored incident response strategy for smart agriculture systems is crucial for adeptly managing security breaches or data privacy issues. This plan should delineate steps for identifying, confining, and addressing security breaches, as well as communicating with stakeholders and regulatory bodies in case of a data breach (Chowdary et al., 2023).

By adopting a holistic strategy encompassing technical safeguards, policy frameworks, and risk management tactics, stakeholders in smart agriculture can mitigate data privacy and security worries. This approach cultivates trust in IoT technologies and guarantees the responsible and secure use of data in agricultural practices.

Infrastructure and Connectivity Limitations in Rural Areas

1. These are characterized by the lack of access to modern technological advancements, reliance on conventional farming methods, and scarcity of publicly available datasets for agricultural research.
2. Challenges in inspecting and maintaining irrigation systems in large rural areas include the installation of sprinklers far apart, making inspection prompt and cost-demanding task. Using technology like the modified MR-CNN (Multi-Region-Convolutional Neural Networks) algorithm aims to automate the detection of water curtains in irrigation systems.
3. Using Zigbee technology for wireless data transmission and the MQTT (Message Queuing Telemetry Transport) protocol for informing field conditions in areas with limited network bandwidth enable remote monitoring and control of irrigation systems, overcoming infrastructure and connectivity limitations.
4. Lack of digital skills and expertise, absence of proper technology infrastructure, and challenges in providing Internet connectivity, standardizing communication protocols, ensuring security, and selecting durable devices hinder the adoption and implementation of AIoT (Artificial intelligence of things) in smart agriculture. (Sharma et al., 2023)
5. Limited internet coverage, poor broadband penetration, impediments to knowledge and education, high IoT system costs, regulatory barriers, and possible employment displacement due to digitalizing agriculture are some of the challenges rural communities face.
6. Infrastructure and connectivity limitations also involve the affordability of commercial solutions, the need for long-range wireless communication, reducing energy consumption, and financial constraints faced by farmers in rural areas. (Kumar et al., 2023).

Cost and Affordability of Smart Irrigation Technologies

1. Cost and Affordability Emphasis: This chapter indirectly addresses the cost and affordability of smart irrigation technologies by highlighting potential cost savings, reduction of water wastage, and economic benefits through automation. It emphasizes the need for cost-effective solutions and affordability considerations.
2. Rising Water Costs: This section explains the need for effective water usage and the rising expense of water. It suggests smart irrigation systems should be implemented using IoT and cloud server technologies to address the cost and affordability issues related to smart irrigation technology.
3. Cost-Effectiveness and Accessibility: Smart irrigation technologies are portrayed as cost-effective and affordable, focusing on reducing water consumption and enhancing productivity, putting into consideration factors like per-capita income and wealth inequality, particularly in the context of India.
4. Barriers and Challenges: This section addresses barriers related to cost factors and challenges in supporting performance and accuracy, indirectly touching on the cost and affordability aspects of smart irrigation technologies.
5. Efficiency Benefits: While not explicitly delving into cost and affordability, the paper underscores the potential benefits and efficiency gains linked to digitization and smart agricultural practices, highlighting their cost-effectiveness.
6. Affordability in Developing Countries: Highlights the significant concern over the cost and affordability of smart irrigation technologies, especially for farmers in developing countries with limited profits, advocates for the use of affordable sensors and controllers to increase accessibility and deal with the financial difficulties that these regions' farmers face (Kumar et al., 2024).

Need for Education and Training for Farmers

Education and training for farmers are crucial in addressing the challenges posed by infrastructure and connectivity limitations in rural areas. By providing farmers with the necessary knowledge and skills, they can effectively use technology and modern agricultural practices to enhance productivity and sustainability. Training programs can empower farmers to utilize IoT devices, data analytics, and automation tools for efficient farm management. Additionally, education initiatives can help bridge the digital divide by improving digital literacy among farmers, enabling them to access and use online resources for market information, weather forecasts, and agricultural best practices. Overall, investing in education and training for farmers is essential for unlocking the full potential of technology in agriculture and overcoming the barriers imposed by infrastructure and connectivity limitations in rural areas (Kumar et al., 2023).

Future Advancements in AI and Their Potential for Irrigation Optimization

The future advancements in AI for irrigation optimization include the following key points:

1. Implementing automated condition sensing devices and intelligent monitoring in agricultural fields.
2. Combining AI-driven models with decision support systems to enhance water supply optimization through transpiration parameters.
3. Employing compact, energy-efficient technologies enhanced by 5G capabilities for remote monitoring and instantaneous data gathering.
4. Employing advanced analytics and interconnected sensors to optimize responses to crop variations, minimizing water wastage while enhancing yields.
5. Highlighting the significance of networked, sensor-driven water monitoring and management for sustainable regulation of irrigation.
6. Further refinements in model inputs and exploration of alternative data pre- and post-processing techniques are needed.
7. Investigation of AI applications in Variable Rate Irrigation (VRI) systems.
8. Advancement of wireless sensor networks and cloud-integrated systems to automate irrigation control, using real-time measurements of substrate water levels.
9. Implementing wireless sensor networks and AI-driven precision irrigation management based on substrate-specific water retention curves in greenhouse soilless cultivation.

10. Emphasis on minimizing water usage, forecasting optimal yields through historical crop and soil analysis, and delivering water to plants at the ideal timing.
11. Enhancing yield by keeping farmers informed about upcoming weather possibilities.
12. Using weather predictions, AI technologies can optimize irrigation schedules, reduce water waste, and improve decision-making.
13. Potential implementation of a short-term weather prediction-based deep Q-learning (DQN) irrigation decision-making system.

These points collectively highlight AI's potential to revolutionize irrigation practices to enhance efficiency, sustainability, and productivity in agriculture.

Conclusion

Incorporating Artificial Intelligence (AI) technologies into smart irrigation systems offers a revolutionary solution to agricultural challenges, particularly optimizing water use and enhancing crop yields. Through the use of advanced algorithms like Convolutional Neural Networks (CNN) and Recurrent Neural Networks (RNN), these systems provide customized irrigation schedules, precise detection of water distribution, and improved decision-making capabilities. The economic and environmental advantages achieved through smart irrigation underscore its potential to encourage sustainable farming practices and tackle significant agricultural issues. Nonetheless, addressing concerns such as data privacy and security, infrastructure constraints, and the necessity for farmer education and training are crucial to fully harnessing the advantages of AI in agriculture. Overall, the fusion of AI and agriculture through smart irrigation systems holds immense potential for transforming water management practices and fostering a more sustainable future for the agricultural sector.

References

Abioye, E. A., Hensel, O., Esau, T. J., Elijah, O., Abidin, M. S. Z., Ayobami, A. S., Yerima, O., & Nasirahmadi, A. 2022. Precision Irrigation Management Using Machine

Learning and Digital Farming Solutions. In *AgriEngineering* (Vol. 4, Issue 1, pp. 70–103). MDPI. https://doi.org/10.3390/agriengineering4010006.

Adli, H. K., Remli, M. A., Wan Salihin Wong, K. N. S., Ismail, N. A., González-Briones, A., Corchado, J. M., & Mohamad, M. S. (2023). Recent Advancements and Challenges of AIoT Application in Smart Agriculture: A Review. In *Sensors* (Vol. 23, Issue 7). MDPI. https://doi.org/10.3390/s23073752.

Albuquerque, C. K. G., Polimante, S., Torre-Neto, A., & Prati, R. C. (2020). Water spray detection for smart irrigation systems with Mask R-CNN and UAV footage. *2020 IEEE International Workshop on Metrology for Agriculture and Forestry, MetroAgriFor 2020 - Proceedings*, 236–240. https://doi.org/10.1109/MetroAgriFor50201.2020.9277542.

Ali, A., Hussain, T., Tantashutikun, N., Hussain, N., & Cocetta, G. (2023). Application of Smart Techniques, Internet of Things and Data Mining for Resource Use Efficient and Sustainable Crop Production. In *Agriculture (Switzerland)* (Vol. 13, Issue 2). MDPI. https://doi.org/10.3390/agriculture13020397.

Altalak, M., Uddin, M. A., Alajmi, A., & Rizg, A. (2022). Smart Agriculture Applications Using Deep Learning Technologies: A Survey. *Applied Sciences (Switzerland)*, *12*(12). https://doi.org/10.3390/app12125919.

Badreldeen, men M., Ragab, M. A., Sedhom, A., Mamdouh, W. M., & Ali Ragab, M. (2022). IoT based Smart Irrigation System. *International Journal of Industry and Sustainable Development (IJISD)*, *3*(1). http://ijisd.journals.ekb.eg.

Chowdary, A. V., Saini N., Kumar A., Kumar S., Ballabh J., Bhatt S. S., Bhatt A., Prakash S., & Patel A. (2023). Influence of integrated nutrient management on physiological parameters of lentil (Lens culinaris Medik.). *Plant Science Today*, 10(3):94–97. https://doi.org/10.14719/pst.2124.

García, L., Parra, L., Jimenez, J. M., Lloret, J., & Lorenz, P. (2020). IoT-based smart irrigation systems: An overview on the recent trends on sensors and iot systems for irrigation in precision agriculture. In *Sensors (Switzerland)* (Vol. 20, Issue 4). MDPI AG. https://doi.org/10.3390/s20041042.

Kanmani, R., Muthulakshmi, S., Subitcha, K. S., Sriranjani, M., Radhapoorani, R., & Suagnya, N. (2021). Modern Irrigation System using Convolutional Neural Network. *2021 7th International Conference on Advanced Computing and Communication Systems, ICACCS 2021*, 592–597. https://doi.org/10.1109/ICACCS51430.2021.9441917.

Kanuru, L., Tyagi, A. K., Aswathy, S. U., Fernandez, T. F., Sreenath, N., Mishra, S. Prediction of pesticides and fertilizers using machine learning and Internet of Things. In *Proceedings of the 2021 International Conference on Computer Communication and Informatics (ICCCI),* Coimbatore, India, 27–29 January 2021; pp. 1–6.

Kumar, A., Das N., Sharma H., & Singh R. Impact of Agrochemicals on the Environment and Ecological Alternatives. In A. Karmaoui (Ed.), Water-Soil-Plant-Animal Nexus in the Era of Climate Change 2024;317-328. IGI Global. https://doi.org/10.4018/978-1-6684-9838-5.ch015.

Kumar, A., Kumar S., Juyal R., Sharma H., Bisht M. Nanoscience in Agricultural Steadiness. In: Dar, G. H., Bhat, R. A., Mehmood, M. A. (eds) Microbiomes for the

Management of Agricultural Sustainability. Springer, Cham. 2023;285-296. https://doi.org/10.1007/978-3-031-32967-8_17.

Kumar, A., Sharma, H., Shakeb, A., & Joshi, M. (2022). Environmental Sustainability: An Emerging Concept. International Year of Millets 2023, 29.

Kumar, R., S. Sachan, R. Kumar, K. Parmar, A. K. Mishra, A. Kumar and A. K. Mishra. 2023. Unveiling Use of Information and Communication Technology (ICT) in Western Uttar Pradesh, India: The Farmers' Perspective. Journal of Global Innovations in Agricultural Sciences 11:521-531. https://doi.org/10.22194/JGIAS/23.1197.

Kumar, R., Singh, A., Yadav, R. B., Kumar, A., Kumar, S., Shahi, U. P., & Singh, A. P. (2017). Growth, development and yield response of rice (Oryza sativa L.) as influenced by efficient nitrogen management under subtropical climatic condition. Journal of Pharmacognosy and Phytochemistry, 6(6S), 791-797.

Kumawat, L., Jat, L., Kumar, A., Yadav, M., Ram, B., & Dudwal, B. L. (2022). Effect of organic nutrient sources on growth, yield attributes and yield of wheat under rice (Oryza sativa L.) wheat (Triticum aestivum L.) cropping system. The Pharma Innovation Journal, 11(2), 1618-1623.

Montesano, F. F., van Iersel, M. W., Boari, F., Cantore, V., D'Amato, G., & Parente, A. (2018). Sensor-based irrigation management of soilless basil using a new smart irrigation system: Effects of set-point on plant physiological responses and crop performance. *Agricultural Water Management*, *203*, 20–29. https://doi.org/10.1016/j.agwat.2018.02.019.

Moysiadis, T., Adamides, G., Stylianou, A., Zotos, N., Giannakopoulou, M., Alexiou, G. Chapter 10 - Use of IoT *Technologies for Irrigation and Plant Protection: The Case for Cypriot Fruits and Vegetables. In Bio-Economy and Agri-Production*; Bochtis, D., Achillas, C., Banias, G., Lampridi, M., Eds.; Academic Press: Cambridge, MA, USA, 2021; pp. 175–194.

Muangprathub, J., Boonnam, N., Kajornkasirat, S., Lekbangpong, N., Wanichsombat, A., & Nillaor, P. (2019). IoT and agriculture data analysis for smart farm. *Computers and Electronics in Agriculture,* 156, 467–474. https://doi.org/10.1016/j.compag.2018.12.011.

Prati, R. C., Polimante, S., K., C., & et al., (2020). Water spray detection for smart irrigation systems with Mask R-CNN and UAV footage. In *Proceedings of the International Conference on Metrology for Agriculture and Forestry* (pp. 1-6). doi: https://doi.rog/10.1109/MetroAgriFor50201.2020.9277542.

Ramachandran, V., Ramalakshmi, R., Kavin, B. P., Hussain, I., Almaliki, A. H., Almaliki, A. A., Elnaggar, A. Y., & Hussein, E. E. (2022). Exploiting IoT and Its Enabled Technologies for Irrigation Needs in Agriculture. In *Water (Switzerland)* (Vol. 14, Issue 5). MDPI. https://doi.org/10.3390/w14050719.

Sami, M., Khan, S. Q., Khurram, M., Farooq, M. U., Anjum, R., Aziz, S., Qureshi, R., & Sadak, F. (2022). A Deep Learning-Based Sensor Modeling for Smart Irrigation System. *Agronomy*, *12*(1). https://doi.org/10.3390/agronomy12010212.

Sharma, H., Kumar A., Singh R, & Kumar S. Impact of Nanopriming and Omics-Based Applications for Sustainable Agriculture. In A. Singh, V. Rajput, K. Ghazaryan, S.

Gupta, & T. Minkina (Eds.), Nanopriming Approach to Sustainable Agriculture 2023;175-191. IGI Global. https://doi.org/10.4018/978-1-6684-7232-3.ch008.

Vallejo-Gómez, D., Osorio, M., & Hincapié, C. A. (2023). Smart Irrigation Systems in Agriculture: A Systematic Review. In *Agronomy* (Vol. 13, Issue 2). MDPI. https://doi.org/10.3390/agronomy13020342.

Valsan, V., Rajesh, K., Santhoshlal, N. M., Pradeep, V. (2022). Smart Irrigation Monitoring System for Multipurpose Solutions. In *Ubiquitous Intelligent Systems*; Springer: Berlin/Heidelberg, Germany, 461–473.

Biographical Sketches

Sharad Sachan

Sharad Sachan is currently working as an Associate Professor and Head of the Department at Lovely Professional University, Punjab, India. He holds a B.Sc. (Hons.) Agriculture and an M.Sc. in Agricultural Economics from Chandra Sekhar Azad University of Agriculture and Technology, Kanpur, U.P. followed by a Ph.D. in Agricultural Economics from Sardar Vallabhbhai Patel University of Technology, Meerut, U.P. With 9.5 years of experience in teaching and research, he has published 21 Scopus-indexed papers, 6 articles, 10 Books and Book chapters and 10 patents. He was also awarded the Excellence in Teaching Award-2019 on 3rd Global Initiative in Agricultural and Applied Sciences for Eco Friendly Environment (GIASE-2019).

Atin Kumar

Atin Kumar is currently working as an Assistant Professor at the School of Agriculture, Uttaranchal University, Dehradun, Uttarakhand, India. He holds a B.Sc. in Agriculture and an M.Sc. in Agriculture with a specialization in Agricultural Chemistry & Soil Science from CCS University, Meerut. He obtained his Ph.D. in Soil Science from Sardar Vallabhbhai Patel University of Agriculture & Technology, Meerut (U.P.) in 2019. He has actively participated in numerous national and international workshops, conferences, and training sessions. His expertise lies in soil fertility, nutrient management, and water quality. With almost 4 years of experience in teaching and research, he has published over 40 research papers, articles, abstracts, and book chapters. Atin Kumar has been honored with the Young Researcher Award, and the Global Scientist Award for his contributions to agriculture and sustainable development.

Nitish Karn

Nitish Karn is currently pursuing in M.Sc. Agriculture (Agronomy) at School of Agriculture, Uttaranchal University Dehradun, Uttarakhand, Inida. He has participated in Global millets (Shree anna) conference 2023 and national conference on Prospects and Challenges: Promoting Sustainable Farming System in Mountain Regions of Uttarakhand in 2024. His expertise lies in soil fertility, water management and nutrient management. He has published over various research papers, articles, abstracts, and book chapters.

Rohit Kumar

Rohit Kumar was born on 08 January 1992 in District Bulandshshr, U.P. He obtained his M.Sc. and Ph.D. in Agronomy from, Sardar Vallabhbhai Patel University of Agriculture & Technology (SVPUA&T), Meerut (U.P.) in 2016 and 2019, respectively. He also qualified for the ICAR/ASRB NET exam in 2017. Rohit Kumar served as an assistant professor at Chandigarh University, Punjab, and SGT University, Gurugram, Haryana, respectively. He has published more than 15 research papers, articles, abstracts, and book chapters. He has also received the "Young Professional Award-2018" and the "Young Scientist Award 2019". His area of expertise is precision water and nutrient management in field crops. With nearly 4 years of experience in teaching, research, consultancy, and agricultural extension, he is also involved in teaching undergraduate and postgraduate students and supervised about 4 M.Sc. Agriculture (Agronomy) students. He is currently working as an Assistant Professor at the Faculty of Agricultural Science, GLA University, Mathura, Uttar Pradesh.

Mohamed A. E. AbdelRahman

Mohamed A. E. AbdelRahman is currently working as an Associate Professor in the Division of Environmental Studies and Land Use (DESLU), the National Authority for Remote Sensing and Space Sciences (NARSS), Egypt. He holds a B.Sc. and M.Sc. degree in Agricultural Sciences from the Faculty of Agriculture, Zagazig University, Egypt in year 2004 and 2009, respectively. He obtained his Ph.D. degree in Soil Science and Agricultural Chemistry from University of Agricultural Sciences (UAS), GKVK, Bangalore, Karnataka, India in the year 2014. He has actively participated in numerous national and international workshops, conferences, and training sessions. His expertise lies in soil survey, land degradation, GIS and remote sensing technology and soil fertility. He is a reviewer and a member of editorial board in various international journals. Additionally, he is a member of various societies and organizations. He has published numerous research papers, articles, abstracts, and book chapters in Scopus and Web of Science indexed journals. Mohamed A. E. AbdelRahman has been awarded with the European Space Agency Prize for the Best Poster at 12th Conference of the African Association of Remote Sensing and the Environment, as well as the Scientific Publication Award from the National Authority for Remote Sensing and Space Sciences.

Chapter 5

Synthesis, Characterization, and Agricultural Applications of Super Adsorbent Polymer Composites

**El-Refaie Kenawy*
and Mohamed M. Azaam**
Polymer Research Group, Department of Chemistry, Faculty of Science,
Tanta University, Tanta, Egypt

Abstract

Two essential elements for crops that restrict agricultural output are fertilizer and water. Society has been driven by exponential population expansion to use more nutrients, particularly nitrogen (N), phosphorus (P), and potassium (K), to increase crop yields. Worldwide, there has been a major problem with decreased fertilizer use efficiency and rising fertilizer demand. This is particularly true for nitrogen (N), the most applied plant nutrient and a major driver of production in agriculture. Urea is one of the most extensively used N fertilizers because of its high N content, inexpensive cost, and simple application. However, urea tends to migrate into the air and aquatic systems by volatilization, runoff, and leaching due to its tiny molecular weight, high solubility, and low thermal stability. These effects bring large financial losses and serious environmental issues, such as contaminated waterways, soil, and oceans, and a decline in farmland biodiversity. Therefore, the primary goals of developing fertilizers are to minimize the risk of environmental pollution, maximize nutrient retention in soil, reduce fertilizer loss, and improve utilization efficiency. Using controlled or slow-release fertilizers is a

* Corresponding Author's Email: ekenawy@science.tanta.edu.eg

In: Irrigation Management
Editors: Maha Elbana and Youssef Brouziyne
ISBN: 979-8-89530-318-4

promising way to minimize environmental effects while cutting the high costs associated with intense application.
Nowadays, efficient water use is important because vast farmland areas, particularly in dry and semiarid regions, suffer from a lack of water resources. Because of their superior water absorption and retention capabilities, super-adsorbent polymers are receiving much attention in this regard as controlled-release, slow-release fertilizers for agricultural applications.
This chapter discusses various preparation methods for some super adsorbent polymer composites and their application in the agricultural field.

Keywords: super adsorbent polymer, composites, agricultural applications, cellulose, chitosan, poly (acrylic acid)

List of Abbreviation

AMPS	2-acrylamido-2-methyl-1-propanesulfonic acid
PEG-$(NH2)_2$	amine-terminated polyethylene glycol
APS	ammonium persulfate
ATP	attapulgite
BTCA	1,2,3,4-butanetetracarboxylic dianhydride
CEEC	carbamoyl ethyl ether of cellulose
CMC	carboxymethyl cellulose
St	cassava starch
CO	castor oil
CS	Chitosan
CHCAUR	chitosan/citric acid/urea
CA	citric acid
CRF	controlled-release fertilizers
P(AA-co-AM)	copolymer of acrylamide and acrylate
DAS	dialdehyde starch
DAP	diammonium phosphate
DAP	Di-ammonium phosphate
D	diffusion coefficient
DN	double network
EFFs	ecologically friendly fertilizers
GO	graphene oxide
HES	hydroxyethyl starch

HEMA	2-hydroxyethyl methacrylate
LGe	lignin
MMT	montmorillonite
MCPA	2-methyl-4-chlorophenoxy acetic acid
MBA	N,N'-methylene-bis-acrylamide
NCNPs	natural char nanoparticles
N	nitrogen
NAE	nitrogen agronomic efficiency
NUE	Nitrogen usage efficiency
P	phosphorus
PAM	polyacrylamide
PAA	polyacrylic acid
SRFs	polymer-based slow-release fertilizers
PVA	polyvinyl alcohol
K	potassium
PACS	(2-pyridyl) acetyl chitosan chloride
RH	rice husk
IPN	semi-interpenetrating polymer
SP	starch polyols
LSMM	starch-based liquid mulch material
s-PHBs	Starch-regulated phosphate hydrogel beads
SCB	sugarcane bagasse
SAH	Superabsorbent hydrogels
TBAF/DMSO	tetrabutylammonium fluoride/dimethyl sulfoxide
TSP	Triple Superphosphate.

Introduction

Two essential elements for crops that restrict agricultural output are fertilizer and water. Society has been driven by exponential population expansion to use more nutrients, particularly nitrogen (N), phosphorus (P), and potassium (K), to increase crop yields. Worldwide, there has been a major problem with decreased fertilizer use efficiency and rising fertilizer demand. This is particularly true for nitrogen (N), the most applied plant nutrient and a major driver of production in agriculture. Urea is one of the most extensively used N fertilizers because of its high N content, inexpensive cost, and simple application. However, urea tends to migrate into the air and aquatic systems by volatilization, runoff, and leaching due to its tiny molecular weight, high solubility, and low thermal stability. These effects bring large financial losses and serious environmental issues, such as contaminated waterways, soil, and oceans, and a decline in farmland biodiversity. Therefore, the primary goals of developing fertilizers are to minimize the risk of environmental pollution, maximize nutrient retention in soil, reduce fertilizer loss, and improve utilization efficiency. Using controlled or slow-release fertilizers is a promising way to minimize environmental effects while cutting the high costs associated with intense application.

Nowadays, efficient water use is important because vast areas of farmland, particularly in dry and semiarid regions, suffer from a lack of water resources. Because of their superior water absorption and retention capabilities, super-absorbent polymers are receiving much attention in this regard as controlled-release, slow-release fertilizers for agricultural applications.

Super-absorbent hydrogels (SAH) are three-dimensional networks that are cross-linked and characterized by their exceptional ability to stabilize a significant amount of water without dissolving. SAH typically have hydrophilic and/or ionic groups bonded to the gel network, promoting a greater degree of hydrogel swelling. While the word "SAH" is sometimes replaced in the literature with "hydrogels," the two terminologies are not synonymous. Compared to normal hydrogels (100%, 1 g g-1), SAHs frequently have a higher water absorption capacity (1000-100,000%, 10–1000 g g-1). In agricultural and horticultural contexts, SAHs have been used to improve the physical properties of the soil and reduce the frequency of irrigation based on these attributes.

SAPs can be divided into two categories based on where they come from natural and synthetic polymers. One clear benefit of SAPs based on natural polymers is their degradability, which includes cellulose, starch, and chitosan. However, because of their slow rate of water absorption, they need to be used more frequently. However, even though SAPs made of synthetic polymers, including polyacrylamide (PAM) and polyacrylic acid (PAA), have a long service life, are inexpensive, and have a high rate of water absorption, their inability to degrade could have a negative impact on the environment and the growth of plants (Wei et al., 2016; Demitri et al., 2013; Chang et al., 2021).

This chapter covers several ways to manufacture some super-adsorbent polymer composites using synthetic and natural polymers, and their uses in the realm of agriculture.

Natural SAPs

Natural SAPs are favorable because of their inherent degradability, biocompatibility, environmental friendliness, and renewability. However, their intricate extraction method makes them more expensive than many synthetic polymers. To improve market prospects, numerous studies have attempted to modify natural polymers with synthetic ones.

Cellulose

Cellulose is the most prevalent natural polymer because it is inexpensive, biodegradable, and biocompatible all at the same time. Neat cellulose has a very low absorbency (less than 30 g g^{-1} in distilled water) because of its extremely thick crystalline structure. Etherification, esterification, and sulfonation of hydroxyl groups are popular methods used to increase water absorbency in cellulose by destroying its tight crystalline structure. An etherification reaction, for instance, was conducted using cellulose and 2-chloroethanol. Following grafting of the resultant hydroxyethyl cellulose onto the copolymer of acrylamide and acrylate (P(AA-co-AM)), the product's maximum water absorbency achieved 240 g g^{-1} as opposed to 200 g g^{-1} of wheat straw-g-P(AA-co-AM), confirming the etherification's effect (Song et al., 2019). Carboxymethyl cellulose was created by etherifying cellulose using chloroacetic acid. Upon grafting acrylic acid (AA) *in situ* on the carboxymethyl cellulose (CMC) backbone using graphene oxide (GO) as a filler, the product (CMC-g-PAA/GO) exhibited optimal water absorbency of 750 g g^{-1} in distilled water and 85 g g^{-1} in 0.9 wt% NaCl solution, respectively (Z. Wang et al., 2017). After cellulose was esterified and cross-linked with 1,2,3,4-butanetetracarboxylic dianhydride (BTCA), a hydrogel was created in the tetrabutylammonium fluoride/dimethyl sulfoxide (TBAF/DMSO) system with a maximum water absorbency of 987 g g^{-1} (Kono, Fujita, and Oeda 2013).

Biopolymers have lately attracted attention in research circles as a potential coating material to mitigate increased operating costs and prevent negative impacts from certain polymer-based slow-release fertilizers (SRFs) that are not biodegradable. The current project aims to create a biodegradable polymer-based sustainable coating material. To this purpose, the Alfa plant was first used as a feasible, sustainable source of lignin (LGe), which was then included in the creation of a three-dimensional cross-linked network using citric acid (CA) as a cross-linking agent and methylcellulose (MC) as a matrix (El Bouchtaoui et al., 2022).

Subsequently, the intended coating material was deposited onto water-soluble fertilizers, specifically Di-ammonium phosphate (DAP) and Triple Superphosphate (TSP), using a rotating pan machine. Additionally, the covering film increased the soil's ability to retain water and strengthened the fertilizer's mechanical resistance. Their extraordinary potential to improve wheat leaf area, chlorophyll content, and biomass, as well as the architecture of the root and the end fruiting efficiency, has also been proven by the agronomic assessment. These findings demonstrated the potential utility of this hybrid composite as an effective coating material for creating slow-release fertilizers with various uses. All prepared films have a swelling ratio that rises with time but eventually reaches equilibrium or its maximum value. It is evident that the methylcellulose swells more quickly in the initial stage before slowing down and achieving an equilibrium water absorbency of 144.4% after around 50 minutes. Methylcellulose is, therefore, regarded as a good water absorbent. Additionally, it is evident that an increase in lignin content positively impacted the swelling behavior of the composite. Furthermore, MCˆ MC85/LGe15 < MC75/LGe25 < MC60/LGe40 was the order in which the swelling rate rose. The MC75/LGe25 and MC85/LGe15 films are responsible for the mild swelling ratios of 174.2 and 182%, which were obtained in roughly 62 minutes, respectively. The film doped with high lignin content recorded the maximum water uptake (225%) in almost 187 minutes. Undoubtedly, lignin increased methylcellulose's water absorbency. This may have happened because lignin's structure contains many free hydroxyl groups that attract water molecules. Compared to MC and the other composite films, the film doped with high lignin content (MC60/LGe40) achieved the equilibrium swelling condition more slowly. In actuality, the MC60/LGe40 swelled to about 46% in the same 20 minutes as the film simply containing MC, which swelled to about 128%. The occurrence of a highly dense and compact structure in the composite surface that prevents water molecules from diffusing quickly to the bulk could account for this phenomenon. After the heating stage, the alcoholic groups of both polymers and the dehydrated citric acid underwent an esterification reaction, which produced the condensed structure that was the outcome of the chemical cross-linking utilizing citric acid. Weight loss was used to measure the MC and MC/LGe films' degradation following a 30-day burial in soil. The findings demonstrated that lignin loadings slowed the MC films' biodegradation. As for uncoated fertilizers and those coated with MC/LGe, the slow-release behavior of phosphate and nitrogen was examined. In the first half hour of soaking in water, over 68.38 percent of the phosphorus and 88.24 percent of the nitrogen from the uncoated DAP were released. Furthermore, within the same time limit, up to 51.96% was liberated from uncoated TSP. The lignin concentration significantly impacts the slow-release process, as demonstrated by the preceding experimental data. Furthermore, it should be noted that the release was significantly delayed towards more extended periods as the amount of lignin loaded increased. Following this, it was also said that the release rate decreases with increasing film density. Due to ester connections between citric acid and cellulose chains, a highly entangled network is formed, resulting in this compaction (Figure 1).

The P complete release times from each form of fertilizer exhibit a minor variation, which may be attributed to the intermolecular interactions between phosphate and other ions inside each granule. In fact, the TSP released nearly 28 times less phosphates, and the DAP coated with a high lignin content released nearly 60 times fewer nutrients than the uncoated one. Consequently, it is essential to note that MC and MC/LGe coating films could significantly postpone the release of nutrients, outperforming other lignin- and MC-based coatings previously documented in the literature.

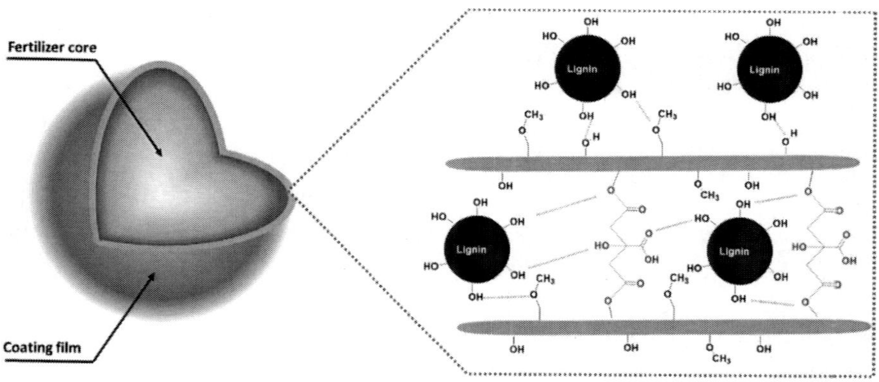

Figure 1. Schematic representation of the chemical structure of the coating film (Adopted from (El Bouchtaoui et al., 2022).

Numerous uses for cellulose and nanocellulose SAH were mentioned in another study (Z. Zhang et al., 2023). A highly appealing product name for specific close-to-use uses of cellulose and nanocellulose SAH is "soil conditioners." By creating double-layered SAH and semi-interpenetrating polymer (IPN) networks, the architectures of SAH have been adjusted to reduce the rate at which fertilizers are released gradually. Semi-IPN could consist of two parts: a linear polymer and a cross-linked polymer (gel network). While double-layered SAH embeds fertilizer in the core layer and releases it as the outer layer degrades, it can also help create a denser and stronger gel network to slow fertilizer release.

The release profiles of cellulose-g-AA in a single network (APP), cellulose-g-AA PVA^{-1}, and semi-IPN (PVA as the second polymeric component) were compared by (W. Wang et al., 2021). They discovered that semi-IPN showed improved wheat growth but a slower rate of APP release.

With urea as the core, ethyl cellulose as the inner coating, and cellulose SAH as the outside coating, (M. Zhang and Yang 2021) created a unique double-coated multifunctional slow-release fertilizer. This improved the soil's ability to store water and produced a slow-release nitrogen profile. Applications for nanocellulose SAH in agriculture have also been reported. When (Barajas-Ledesma et al., 2020) examined the effect of TEMPO-oxidized nanocellulose on radish germination in Petri dishes, they discovered that nanocellulose SAH increases the germination index by 30–40%.

In a further study, commercial SAP and TEMPO-oxidized nanocellulose SAH were used as soil conditioners for spinach in sandy and clay loam soils (Barajas-Ledesma et al., 2022). It is interesting to see that those plants that were given nanocellulose SAH showed slower development. The high degradability of nanocellulose may have caused the targeted soil amendment capability to malfunction. This discovery also clarified why, although cellulose is a minor component, cellulose-graft-polymer has been the most often used soil conditioner in cellulose SAH until date (degradation of cellulose did not impair the overall soil amendment quality). (Capanema et al., 2018) created a superb SAH by synthesizing carboxymethyl cellulose-PEG SAH with the organic crosslinker citric acid, which exhibits encouraging characteristics of biocompatibility and biodegradability. The resulting hydrogel's water adsorption behavior showed that its swelling degree usually ranged from 100% to 5000%. Citric acid was determined to be the primary cause of this super absorbency because it increases cross-linking and keeps the network's negative ends bonded to citric acid, which would otherwise reveal the network's hydrophobic nature (Figure 2).

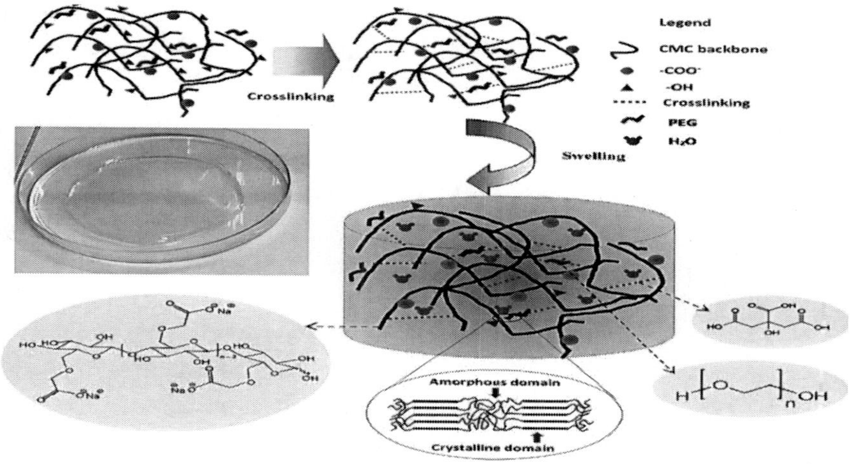

Figure 2. Representation of the cross-linked hydrogel network. Swollen structure after immersion in water (Adopted from Capanema et al., 2018).

Starch

Because of its availability, biodegradability, and environmentally beneficial qualities, starch is a renewable natural polysaccharide biopolymer with great promise. The starch-based hydrogel can potentially increase the soil's ability to hold water significantly because of its network structure, hydrophilic nature, and capacity to create strong hydrogen bonds with water molecules. As a result, soils fertilized with these hydrogel fertilizers would retain more water. Starch is a biomacromolecule derived from glucose and commonly utilized as a water-retaining agent. It possesses excellent hydrophilicity, degradability, and environmental friendliness, but its water absorbency, mechanical stability, and salt tolerance are weak, with levels below 10 g g^{-1} in distilled water. One popular technique for increasing the water absorbency of starch-based water retention agents is grafting AA to starch.

For instance, AA was utilized in solution polymerization with starch treated with sulfamic acid. Because the sulfonic acid groups adsorb and transfer water molecules, the final product's water absorbency increased by 100 and 13%, respectively, over starch-grafted acrylic polymer. It was 1026 g g^{-1} in deionized water and 145 g g^{-1} in 0.9 wt% NaCl solution (Zhao et al., 2019).

A solution of 2-acrylamido-2-methyl-1-propanesulfonic acid (AMPS) was prepared by mixing AA with cassava starch (St) to increase the cross-linking density and storage modulus of St-g-AA-AMPS SAP. According to the results, implementing AMPS helped form the ideal network structure. In distilled water and 0.9 weight percent NaCl solution, the maximal swelling ratios were 1200 g g^{-1} and 90 g g^{-1}, respectively. AMPS' hydrophilic ability and high ionization constant significantly enhanced the re-swelling capability, water retention, and salt tolerance (Meng and Ye 2017).

Hybridizing starch and inorganic fillers is another strategy to increase water absorption even in saline solutions and under load. For instance, compared to 286 g g^{-1} of plain hydrogel, the water absorbency of a nanocomposite hydrogel made of starch-g-(P(AA-co-AM))/polyvinyl alcohol and nano-clinoptilolite increased to 365 g g^{-1} in distilled water. Additionally, it became potentially useful for agricultural application because the water absorbency in 0.9 wt% NaCl solution increased from 46 g g^{-1} of pure hydrogel to 54 g g^{-1} of nanocomposite hydrogel, and the water absorbency at 0.9 psi increased from 30 g g^{-1} of neat hydrogel to 37 g g^{-1} of nanocomposite hydrogel (Olad, Doustdar, and Gharekhani 2018).

Furthermore, compared to neat hydrogel (202 g g^{-1} in distilled water), the addition of chemically modified natural char nanoparticles (NCNPs) to the starch-g-P(AA-co-AM) super-absorbent demonstrated a significantly increased water absorbency (390 g g^{-1} in distilled water). This is likely due to the coexistence of the physical and covalent cross-linkage in the NCNPs/Hydrogel nanocomposite. After 14 days, the nanocomposite hydrogel's water-retention (%) capacity rose from 23% to 7% of the plain hydrogel. Also, there was an improvement in the absorbency of NCNPs/Hydrogel under load (Motamedi et al., 2020).

The review provides significant insights into the potential of starch-based formulations to enhance nutrient management, increase crop yield, and promote sustainable agriculture. It compiles current research on starch, its modifications, and its applications in SRFs (Salimi et al., 2023).

Designed to extend the bioavailability of nutrients through crop growth cycles, ideally with a single application, slow-release fertilizers (SRF) and controlled-release fertilizers (CRF) minimize nutrient losses and root damage in comparison to conventional methods (Mikula et al., 2020), thereby increasing fertilizer use efficiency (Lawrencia et al., 2021).

Three techniques are used to manufacture slow-release fertilizers: Due to their ease of production, matrix-type formulations account for the majority of SRFs. The active component can spread throughout a porous matrix thanks to the carrier phase's pores (Elhassani et al., 2019). Coated fertilizers, which contain an inert material covering the fertilizer's core, comprise the second major class of SRFs. The release of fertilizer nutrients is regulated by diffusion via a coat (El Gharrak et al., 2022). The third main class of SRFs is chemically controlled release products, such as urea-formaldehyde and polyphosphates (G. Yang et al., 2023).

Starch can be altered in several ways to improve its effectiveness as a suitable coating. Specific modification techniques, like etherification or esterification, can make the starch more hydrophobic, which lowers nutrient leaching and enhances controlled release qualities. These techniques effectively reduce nutrient losses and ensure that plants have a steady supply of nutrients. On the other hand, grafting methods can strengthen and stabilize starch, enhancing water absorption and regulating nutrient release (Qiao et al., 2016).

Numerous studies on using starch or modified starch in the formulation of slow-release fertilizers have been published in the literature.

(Xiao et al., 2017) synthesized starch-based super-absorbent polymers (SBSAPs) with slow-release qualities using a one-step melt blending approach utilizing starch as a renewable resource. Analyzing the amount of urea released in water, it was discovered that less than 15% was released in a day and more than 80% after 30 days. The profile of urea release showed different phases. Due to dynamic water exchange, very little urea was released in the early stages. As the gel strength dropped in the middle stage, a sizable amount of urea was released more quickly. Eventually, as the material degraded and decomposed during the later period, the majority of the urea was released. It is possible that some urea was integrated with the unfrozen water in the SBSAP, preventing it from being discharged. The substance had broken down and considerably reduced its strength after the 45-day release period. Overall, the results highlight the urea release behavior from the SBSAP composite and emphasize the significance of the dynamic interaction between the composite and water and gel strength (Xiao et al., 2017).

In another work, starch was grafted with acrylic acid (AA) in the presence of urea to generate urea CRF. According to the results, using super-adsorbent hydrogels derived from starch and acrylic acid on urea fertilizer enhanced soil water-holding capacity and decreased urea release. The soil sample with the highest water-holding capacity, which contained 0.25 percent hydrogel super-adsorbent, had a capacity of 120%. The cross-linking density of the hydrogel influenced its releasing performance. Just 25% of the urea was released in the early stages (1–5 days) when the polymeric network was robust. After that, more than 64% of the urea was released throughout the 5–20 day interval and more than 90–99% during the 20–30 day period. The hydrogel's gel strength weakened with increasing exposure time, which caused the three-dimensional network to weaken and partially disintegrate, allowing the residual urea to escape. The hydrogel's presence significantly altered the soil's chemical and physical characteristics, which in turn affected the plants' ability to thrive in the experiment. In germination tests, seeds planted in urea-encapsulated hydrogel-treated soil produced shoots that were 2.8 cm long, but seeds planted in untreated soil produced shoots that were 0.5 cm long. Given that urea is essential for plant growth, the hydrogel's presence encouraged the growth of the plants. Nevertheless, it is important to remember that too much water in the soil can hinder germination until it evaporates or is naturally absorbed, particularly in light of the chosen soil's low water-holding capacity. By creating a favorable environment, the hydrogel-treated soil was a suitable medium for plant growth and seed germination. Similar trends in plant height were noted after 15 days, with heights of 5 cm and 1 cm for soil containing urea-encapsulated hydrogel and SAH 2, respectively, while untreated soil did not support any plant development at all because of the dry circumstances brought on by water vaporization (Sarmah and Karak 2020).

Starch-regulated phosphate hydrogel beads (s-PHBs), which are polysaccharide supramolecular hydrogel beads made of cassava starch and sodium alginate via Ca^{2+} cross-linking gelation, are made using a straightforward method recently presented by (Tiamwong et al., 2023) The findings demonstrated that phosphate release from PHBs was more influenced by solution pH than by s-PHBs. The greater the pH of the fluid rose, the more phosphate that PHBs released. Phosphate release increased by almost 50% when the solution pH was raised from pH = 5 to pH = 9, most likely due to structural alterations. There are various benefits to adding starch to ecologically friendly fertilizers (EFFs) to save fertilizer prices. First off, starch is a cheap and widely available natural polymer. Second, adding starch to EFFs makes them more chemically and physically stable. This boosts effectiveness and lowers losses. This increased efficiency can result in a decrease in the quantity of fertilizer needed to produce the appropriate crop yields. Lastly, plants can absorb nutrients more effectively thanks to the regulated release characteristics of starch-based EFFs (Tiamwong et al., 2023).

PVA, or polyvinyl alcohol, is a biodegradable plastic. It can enhance the coating characteristics when mixed with starch to create slow-release fertilizers. The mechanical characteristics of the coating film are also enhanced by adding clay nanoparticles. (Sarkar et al., 2021) used polyvinyl alcohol, bentonite clay, wheat starch, etc., to encapsulate and regulate the release of diammonium phosphate (DAP). First, a gel was created by combining starch and water at 90°C; next, water and polyvinyl alcohol were combined in a different container, and the temperature was progressively raised to 90°C. The water-based PVA and starch solutions were combined using a magnetic stirrer/heating unit and placed in a four-necked borosilicate glass vessel after cooling to 30ºC. An ammonium persulfate (APS) solution was added, and the temperature was elevated to 70ºC to start the radical chain reaction, which produced a water-based activated starch/PVA combination. In multiple batches, clay-fractionated bentonite was added at 0, 5, 10, 15, and 20 weight percent (wt%) to the water-based activated starch/PVA blend that was formed. The mixture was constantly swirled. The resulting cross-linked filmogenic starch/PVA/bentonite was treated with formaldehyde. For full polymerization, the mixture was mixed for three hours at 90 °C (Sarkar et al., 2021).

(Salimi et al., 2021) created an aqueous polymeric covering of starch-g-poly (styrene-co-butyl acrylate) latex enhanced with natural char nanoparticles (NCNPs) to create a slow-release fertilizer. Free radical polymerization was utilized to create starch-based nano-composite latexes, which were then grafted onto the starch backbone in the presence of NCNPs using vinyl monomers (styrene and butyl acrylate). The results showed that incorporating NCNPs into coating formulation greatly improved plant growth and slowed the rate of N leakage. The efficiency of all the synthesized SRF samples was higher than that of the SBR (poly(styrene-co-butyl acrylate)-coated urea), indicating the starch's critical role in the coating formulations. The percentage of total nitrogen and nitrate content in the tomato plants' aerial sections was higher (21.1%) and lower (42.1%) in the sample containing 1 w% NCNPS (F13) than in the pure urea. Nitrogen usage efficiency (NUE), nitrogen agronomic efficiency (NAE), and nitrogen apparent recovery fraction (NRF) may all be increased by using the F13 formulation in particular. These metrics are higher than those of uncoated urea, at 68.3%, 49.9%, and 29.7%, respectively.

To overcome initial burst release (PCS-Fe-P/zein) and create a kind of SRF, (Qi et al., 2020) loaded phosphate and iron ions onto carboxymethyl starch and used zein as a renewable, readily available, and reasonably priced coating material. SEM pictures abundantly showed that the carboxymethyl starch surface was highly porous. However, porosity was decreased by iron absorption, while phosphorus absorption resulted in the formation of a goose egg-shaped structure. After 30 days, the predicted phosphorus release rates from PCS-Fe-P and PCS-Fe-P/zein were 18% and 28%, respectively. Only

10% of the nutrients PCS-Fe-P and PCS-Fe-P/zein were released on the fourth day, but 100% of MAP was released.

(Tian et al., 2021) controlled the release of urea granules using polyurethane, castor oil, and starch. They altered castor oil (CO) and starch polyols (SP) through *in situ* polymerization. Urea was pumped into the rotating drum mechanism to coat the fertilizer granules. After applying polyolefin wax to the fertilizer granules as a surface modification, the wax-coated urea was sprayed with a mixture of SP, CO, and PAPI. The findings demonstrated that because starch polyol had a higher surface porosity, it enhanced the release of nitrogen. Because castor oil is not very hydrophilic, it produced a thick layer that inhibited the release of nitrogen. Castor oil controlled nitrogen release and starch polyol mutual modification for a maximum of seven months. After nine months, the rates of deterioration of petroleum-based polyurethane and modified starch and castor oil polyol coatings were 3.74% and 5.05%, respectively. The PAPI ratio, CO ratio, and coating % substantially impacted the N-release characteristics of BCPCU. Longer N release times were obtained by increasing the PAPI ratio, whereas N was released through the coated shell more quickly when the PAPI ratio was low. Higher CO ratios indicate slower nutrient delivery. The CO ratio influenced the initial release rates and prolonged the N release time. The MDA content of the rice seeds suggested that the coating ingredients during the cultivation process caused no harmful stress. The majority of the compounds dissolved from the coating materials were long-chain alkanes, which are generally safe for plants, according to GC/MS analysis. On the other hand, BCPU showed signs of certain potentially harmful materials, indicating that the -NCO group's reaction to film formation was incomplete. To stop dangerous compounds from leaking out, the polyol ratio can be optimized to achieve the full reaction. Furthermore, the existence of a component in PU generated questions about how it would affect microbes; however, the risk might be reduced by substituting safe dyes. Low-concentration coating materials did not negatively impact plant growth, according to germination studies, and any possible risk from the compounds could be mitigated by using safe dyes and optimizing the coating ratio.

Polyacrylic acid (PAA) was grafted onto starch to create a high-performance starch-based liquid mulch material (LSMM) by Chen et al., The characteristics of LSMM films were greatly enhanced by PAA grafting onto starch. The starch molecules' crystallinity was decreased, and hydrogen bonds were broken by the PAA, resulting in forming a compact structure in the LSSM films. Consequently, the LSMM demonstrated exceptional relative hygroscopicity, water retention, degradability (weight loss of 72.61%), and growth and germination rate promotion (84.00%) in lettuce (Chen et al., 2020).

Utilizing NPK-SRF formulations made of a combination of starch-based latex, hydrogel nanocomposite, and carnauba wax emulsion, the research team recently observed an improvement in tomato output and quality (Motamedi, Safari, and Salimi, 2023). First, three water-based bio-polymeric formulations were created: an emulsion of carnauba wax, a starch-g-poly (styrene-co-butylacrylate) latex, and a starch-g-poly (acrylic acid-co-acrylamide) nanocomposite hydrogel. Next, various ratios of latex and wax emulsion (R-treatment) were used to coat fertilizers (urea, potassium sulphate, and superphosphate granules). Additionally, fertilizers containing nanocomposite hydrogel, called the D and H treatments, were substituted for portions of the coated fertilizers (15 and 30 weight percent, respectively). Lastly, the effects of the as-prepared NPK-SRFs on tomato growth at two distinct levels (100 and 60) were contrasted with commercial fertilizers (NPK treatment) and a commercial SRF (T treatment). According to the results, H100 produced the highest yield (1671.54 g), highest dry matter percentage (9.52%), and highest agricultural agronomy efficiency of fertilizer. Furthermore, H100 was shown to have the highest levels of vitamin C, antioxidant capability, and lycopene. The synthesized SRF samples showed a considerable decrease in nitrate accumulation in tomato fruit compared to NPK100. H100 showed the lowest amount of nitrate accumulation, at 55.24% less than NPK100.

A biodegradable Schiff-base hydrogel urea was synthesized with significant water retention and some slow-release capability. First, potato starch and polyethylene glycol were used to create dialdehyde starch (DAS) and amine-terminated polyethylene glycol (PEG-$(NH_2)_2$). Then, using an *in-situ* reaction between the amino group of PEG-$(NH_2)_2$ and the aldehyde group of DAS, a new Schiff-base hydrogel (SH) was created (W. Zhang et al., 2024). As seen in Figure 3, PEG-$(NH_2)_2$ is altered by PEG activation. Before the amination process, PEG-OTS is synthesized as a precursor. Tosylate groups make excellent candidates for substitution processes that facilitate the conversion of various functional groups. When the concentration of sodium periodate is changed, potato starch is oxidized, resulting in the conversion of C-2 and C-3 hydroxyl groups into aldehyde groups. The amino group of PEG-$(NH_2)_2$ and the aldehyde group of DAS underwent a Schiff-base reaction.

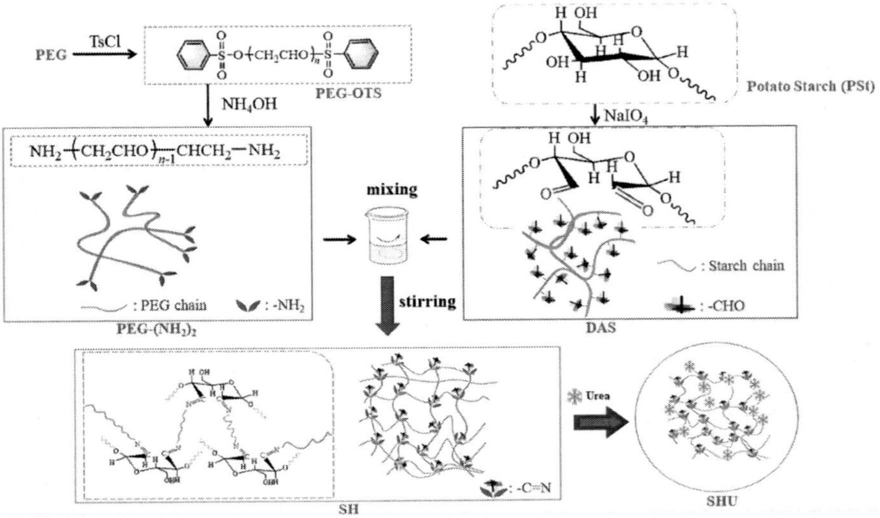

Figure 3. The Schematic diagram for preparing PEG-(NH2)2, DAS, SH and SHU (Adopted from (W. Zhang et al., 2024).

Using SH as the substrate, three SH-based slow-release urea were produced. These are known as SHU1, SHU2, and SHU3, and they differ in terms of urea concentration. The SEM results showed that SH had a maximum water absorption capacity of 4440% ± 6.23% and a porous three-dimensional network structure. After 30 days of release in soil, SHU performed better in terms of slow-release than pure urea; SHU1 had a residual nitrogen level of precisely 36.01 ± 0.57% of the initial nitrogen content. A pot experiment with pakchoi confirmed SHU's ability to retain water and promote plant development. The presence of PEG and their potent interaction forces may increase the starch's ability to absorb water and improve its mechanical characteristics, enabling the creation of a composite of PEG and potato starch through chemical blending. More water molecules can permeate into the three-dimensional hydrogel network through a bigger pore size of SH, which raises the expansion ratio and increases water retention. When SH3's swelling behavior in distilled water was evaluated at room temperature, it was discovered to have a remarkable 4440% ± 6.23% water absorption capability. Following five iterations of dehydration and water absorption, a declining tendency was seen in the hydrogels. Following five cycles, the water absorption capacity decreases by 77%, but the water absorption rate stays at 1035 percent. The breakdown of internal physical cross-linking points during dehydration and water absorption was the cause of the hydrogel's decreased swelling capacity after five cycles, which also led to a weakening of mesh entanglement. As the bulk of the hydrogel increases, so does its ability to retain water in soil. Compared to 0.5 g, 1.5 g hydrogel can retain 44% more water. To assess biodegradability, three created materials—PEG-$(NH_2)_2$, DAS, and SH—were buried in the soil, and their mass loss was calculated at various intervals. Over time, all samples gradually deteriorate. PEG-$(NH_2)_2$, DAS, and SH were shown to have degradation rates of 46.63%, 57.21%, and 36.56% at 35 days, respectively, demonstrating their exceptional degradability.

In a different article, calcium alginate (CA33) was utilized as the primary carrier for regulating and controlling soil water storage, and hydroxyethyl starch (HES) and 2-Acrylamide-2-methyl-1-propanesulfonic acid (AMPS) as raw materials were used to prepare a novel type of super-adsorbent HES-PAM/bentonite and HES/CA33-PAM/sodium bentonite composite materials using *in-situ* polymerization. The *in-situ* polymerization of hydroxyethyl starch and calcium alginate to prepare the super-adsorbent hydrogel is illustrated in Figure 4 (Xiong et al., 2022). According to the results, the novel super-adsorbent polymer achieved salt equilibrium in less than 30 minutes. In 0.9 weight percent saline solution, tap water, and distilled water, the water absorption rates were 1484, 312, and 121 g g^{-1}, respectively. The water absorption rate of the super-adsorbent materials in distilled water and saline solution increased by 25.8% and 10%, respectively, as a result of the effective inarching of calcium alginate, which also produced a gradual water release and enhanced the soil's ability to hold water (Xiong et al., 2022).

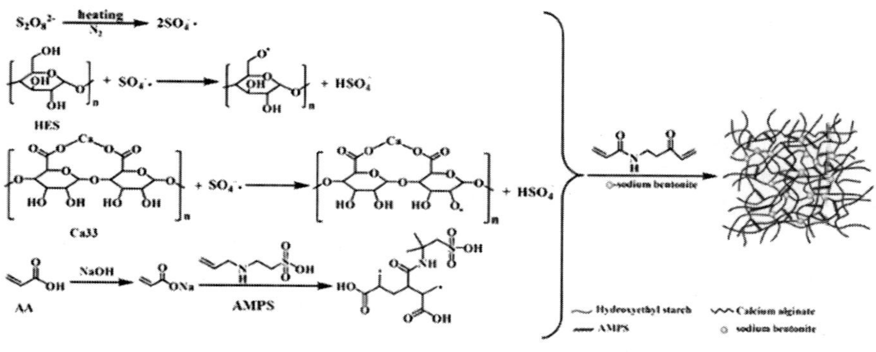

Figure 4. *In-situ* polymerization of hydroxyethyl starch and calcium alginate to prepare the super-adsorbent hydrogel (Adopted from (Xiong et al., 2022).

The hydrogel's water absorption rate increases dramatically with the addition of calcium alginate, and it increases by 25.8% and 10%, respectively, in distilled water and 0.9 weight percent NaCl solution. The composite material exhibits optimal anti-cracking and water evaporation inhibition at 0.10% hydrogel addition. The water absorption and retention capacity in the saline solution were improved, and the evaporation rate was 9.14% lower than that of pure soil, reaching 7.87% after 120 hours. The information mentioned above demonstrates that the synthetic material, when mixed with soil, can absorb and hold large amounts of water. It can also be used to restore and develop saline-alkali land or agriculture, prevent water evaporation and movement beneath the root zone, and release water gradually as needed by the plants. This feature is especially useful for crops grown in soil with an excessively high salt content.

The discussed information will be very helpful in designing bio-friendly and low-cost SAHs for biomedical and agricultural purposes (Qureshi et al., 2020). Hybrid hydrogels based on pre-gelatinized starch containing both natural and synthetic polymers are made using a two-step process. In distilled water and saline solutions, absorbency is significantly increased by surface cross-linking caused by microwave and conventional heating. This surface change aimed to address the hybrid hydrogels' low gel strength. They are excellent candidates for industrial applications due to their high structural contribution of natural components (64%), low cost of pre-gelatinized starch, and moderate and acceptable absorbency value (Ashkani et al., 2018).

According to (Fekete et al., 2017), starch/CMC SAHs may provide a more affordable and effective option than pure cellulose derivative-based gels, depending on the application. Starch and carboxymethylcellulose (CMC) are combined in aqueous solutions to form Ɣ-irradiated starch/CMC SAHs. A high solute concentration and starch supplementation partially replaced the CMC, increasing both the water absorption and gel fraction. A heterogeneous gel structure consisting of scattered starch pieces and granules inside the CMC matrix. After adding starch to the CMC matrix, higher water uptake at high electrolyte concentrations was observed.

Chitosan

Chitosan (CS) is the second most common polysaccharide after cellulose. These days, due to its non-toxicity, antibacterial qualities, biocompatibility, and biodegradability, it is acknowledged as a significant biomaterial. Because of its cationic character, CS can form multilayered networks with other naturally occurring or artificially generated negatively charged polymers. Excellent biocompatibility, biodegradability, and repeatability are some of the inherent qualities that make chitosan a highly desirable material for water retention agents. However, chitosan's water absorption capacity is appallingly low - less than 10 g g^{-1}. Benefiting from a high concentration of hydroxyl and amino groups, chitosan offers a chance to enhance its water absorption through chemical modification (Anitha et al., 2014).

Numerous strategies have been devised to increase chitosan's water absorption capacity. For instance, in an aqueous solution, (2-pyridyl) acetyl chitosan chloride (PACS) was selected to polymerize with AA and AM. The resulting super-adsorbent polymer demonstrated exceptional water absorbency, with a capacity to absorb 615 times its dry weight of distilled water and 44 times its dry weight of 0.9 wt% NaCl solution (Fang et al., 2019).

Furthermore, given that the amino group serves as the primary grafting site in the reaction of AA-grafted chitosan, it was hypothesized that adding more amino groups to the chitosan backbone would enhance the chitosan-g-AA's ability to absorb water and withstand salt. To graft acrylic acid, 2-chloroethylamine hydrochloride-modified chitosan was used. The findings supported the notion that, under ideal circumstances, the product's water absorbency was 644 g g^{-1} in water and 99 g g^{-1} in 0.9 wt% NaCl solution (Fang et al., 2018).

Developing and implementing innovative approaches is crucial to efficiently administer synthetic pesticides (SPTCs) at lower doses while maintaining the requisite efficacy. In addition, to rid the environment of pollutants, these SPTCs must be removed. This requires both effective SPTC detection and the development of substitute plans for the SPTCs' original intended use. Many types of materials, including nanomaterials, are developed to satisfy the abovementioned requirements. One of these nanomaterials is Chitosan-based nanomaterials/nanoparticles (CNPs), which have attracted much interest (Kandasamy, Manisekaran, and Arthikala, 2024).

In addition, because of its advantageous qualities, chitosan is used in all necessary agricultural applications, including the removal, regulated delivery, and sensing of PTCs. Hydrogels are networks of polymers joined by numerous cross-links that can hold onto water and exhibit swelling properties. Hydrogels composed of natural polymers, such as chitosan, which is known for its excellent gelation qualities, can also be used to get rid of herbicides.

For example, chitosan-based hydrogels are focused on removing HRBCs butachlor, a selective chloroacetamide that is toxic to humans and aquatic life but is employed to control weeds and grass during the growth of wheat and rice crops (Kaur, Goyal, and Agnihotri 2021). Here, the hydrogels are created by repeatedly freezing and thawing a nanocomposite containing 2.5% chitosan and 3% poly (vinyl alcohol) and then connecting the two using 50 mM glutaraldehyde. At the lowest pH of 3, the hydrogels effectively removed up to 86.55% of butachlor; however, after adding silver nanoparticles to the hydrogels, the removal of butachlor was increased to 96.87%. This increase in butachlor removal was attributed to increased surface area, improved thermal stability, and various interactions, including hydrogen bonding, electrostatic interactions, and dipole-dipole interactions. However, the removal percentage for non-silver and silver-incorporated hydrogels has dropped to 81.17% and 90.52%, respectively, with a pH increase from 3 to 11.

In a different study, beads of silver nanoparticles cross-linked to chitosan through epichlorohydrin were inserted into a water filtration system column. The beads effectively eliminated an HRBC, atrazine, a chlorinated high-toxic triazine, with a total adsorption capacity of 380 μg ml^{-1} (Saifuddin et al., 2011).

Similar to this, (Dwivedi et al., 2014) removed Opp-INSC and methyl-parathion by combining gold nanoparticles (0.5 wt%) with chitosan in the form of hydrogel beads (with approximately 96% water and stable until approximately 200ºC). In this study, chitosan served as a binding matrix for hydrogel beads that were subsequently created through pH-induced coagulation and as a stabilizer during the synthesis of gold nanoparticles. The removal of methyl-parathion from gold nanoparticle-loaded beads increased from 16 to 55% when the adsorbent dosage was increased from 2 to 12 g l^{-1}. This is greater than the 1 to 12% increase for non-gold-based beads and may be primarily attributed to the gold nanoparticles' effective sulfur (in the adsorbate) and amine group (in the adsorbent) interactions.

Permethrin is an INSC based on pyrethroids, and the Dehaghi group has studied its elimination using silver oxide nanoparticles embedded in chitosan beads (Rahmanifar and Moradi Dehaghi 2014). In a pH of 7, these beads, at a dose range of 0.5-1.5 g, have eliminated up to 98% of permethrin (at 0.1 mg l^{-1}). Under comparable experimental conditions, the group successfully separated nearly 99 percent of permethrin using chitosan beads loaded with zinc oxide nanoparticles. They have also demonstrated that 56% of the beads can be reused even after three cycles.

Recently, Vieira et al., prepared MCNs-based hydrogels using acrylic acid (for carboxyl group generation), ammonium persulfate (as an initiator), and methylene bis acrylamide (as a cross-linker) for the adsorption of 2,4-dichlorophenoxyacetic acid (a selective HRBC to control broad-leaf weeds and has low-to-moderate toxicity under different forms like salt, ester, and acidic) (Vieira, Becegato, and Paulino 2021). A respectable 18.06 ± 0.16 mg of HRBCs have been adsorbed by the MCNs hydrogels; this is marginally less than the 19.01 ± 0.30 mg of HRBCs adsorbed on empty chitosan hydrogels. This may be explained by the fact that the herbicide is prevented from adhering to the chitosan hydrogel's active sites (such as amine and carboxyl groups) by the first adsorption of nanoparticles. Furthermore, an increase in pH from 2 to 5 has decreased the adsorption capacity for both empty and MCN hydrogel. The HRBC's adsorption kinetics have been best fit by a pseudo-second-order kinetic model, which achieves a maximum removal of 90% from real-time distilled and fortified lake water samples.

Another study of a similar nature found that only chitosan-based beads could remove a maximum of ~99% of glyphosate HRBC by increasing the adsorbent dosage (i.e., up to 0.5 g) and contact time (240 min). The adsorbate's adsorption behavior was correlated using models such as the pseudo-second order kinetics model and the Freundlich isotherm (Saelim et al., 2020).

A study has been conducted to provide an update and discussion of significant facets of the synthesis, characterization, and uses of synthetic aniline hydrogels (SAHs) based on significant natural polymers and their use in many domains (Rather, Bhat, and Shalla, 2022).

Using a hydrothermal method, chitosan, citric acid, and urea were combined to create chitosan/citric acid/urea (CHCAUR) SAH (Narayanan et al., 2018). Comprehensive research was done on the effects of various contents, such as greater electrostatic attraction and the presence of macropores, on swelling. The increase in electrostatic contact is the primary cause of the increase in cross-linking brought on by citric acid protonating chitosan. When urea and citric acid mix, gases like carbon dioxide and ammonia are released, which causes holes to form. The produced hydrogel, according to the authors, can absorb 210 $g\ g^{-1}$ of 0.1% NaCl solution and 1250 $g\ g^{-1}$ of distilled water. Additionally, it was shown that the produced hydrogel, even in small amounts, retains water for a long period, lowers the pace at which water evaporates, and functions as a promising all-purpose soil supplement.

Amine ethyl chitosan and acrylic acid were combined using free radical polymerization to create an SAH based on chitosan-acrylic acid. When operating under ideal conditions, it was discovered that the produced sample has a water absorption capacity of up to 644 $g\ g^{-1}$. The presence of COO^- groups, which raise osmotic pressure, causes this much absorption (Fang et al., 2018).

Ecofriendly Chitosan-natural Cellulose cross-linked SAH was created using two fully biodegradable polymers: cellulosic fibers cross-linked with carboxymethylated chitosan. The sample exhibits retention values for water and saline solutions of 610 and 85 g g^{-1} gel, respectively. The additional carboxyl and hydrophilic groups of cellulose and hemicellulose cause the increased absorbance in water. Its high pore architecture, which has open pore geometry and pore sizes between 350 and 600 um, is another factor contributing to its extreme absorbency. This feature enables quick solute penetration into the hydrogel matrix. The reason for the low swelling in saline solution is the salt cations' charge screening effect, which lowers anion-anion repulsions and causes the osmotic pressure to drop (Alam and Christopher 2018).

Alginate

As a naturally occurring SAP with a large number of hydrophilic carboxyl groups, alginate has the potential to be used as a water-retention agent. Brown algae is the source of sodium alginate, a naturally occurring polymeric polysaccharide. It comprises two monosaccharides: α-L-guluronic acid (G) and β-D-mannuronic acid (M). Alginate's macromolecular structure includes these two monomers joined in a straight line. The scientific community is increasingly interested in creating alginate-based composites to address drought and nutrient loss in agriculture and promote sustainable growth. It performs well regarding adhesion, compatibility, and biodegradability, which is perfect for regulating the rate at which nutrients are released. Finding a method that can release fertilizer steadily is very crucial to minimize the loss of urea or other nutrients in agricultural soil (N. Wang et al., 2023). The excellent adsorption capacity of sodium alginate creates new opportunities for developing slow-release fertilizers (Vu and Wu, 2022). A super absorbent nanocomposite can be created by copolymerizing alginate with a polymer to increase the swelling capacity of alginate gels.

For instance, (Olad et al., 2018) created a semi-interpenetrating polymer super-adsorbent nanocomposite (semi-IPN) using sodium alginate-g-polyacrylic acid/polyvinylpyrrolidone and montmorillonite. The semi-IPN's equilibrium swelling capacity increased from 521.17 g g^{-1} to 618.92 g g^{-1} compared to the pure gel, contributing to the nanocomposite's better network structure. Furthermore, the kind, concentration, and pH of the salt solution all have an impact on the alginate gels' ability to swell (Su and Chen, 2018).

Alginate-based composites have been widely used in soil amendment and regulation due to their ability to improve soil water utilization efficiency and decrease fertilizer loss while also having non-toxic, biodegradable, and environmentally friendly properties.

Alginate-based composites have mostly been used in agriculture for insect and pesticide control, aquasorb, slow-release fertilizer, soil amendment, and microbial inoculants. Aquasorb's excellent water-retaining ability, non-irritating qualities, and high-water absorption multiplicity have made it a popular choice in agriculture. It has become another important agrochemical after mulching films, fertilizers, and pesticides (S. S. Li et al., 2023). Like aquasorb, alginate possesses abundant hydrophilic carboxyl and hydroxyl groups and the benefits of a natural polymer. They are becoming increasingly popular because, after cross-linking with other materials, they can significantly increase water retention capacity (D. Yang, 2022).

Ca^{2+} cross-linking alginate nanoparticles, for example, have been synthesized, and their potential for sustainable water release in agricultural areas has been assessed. According to this study, untreated soil lost almost all of its water in just seven days, whereas soil combined with alginate nanoparticles kept about 20% of its water even after 11 days. This is attributed to the improved water absorption of nanoparticles following the $CaCl_2$-induced cross-linking of sodium alginate. Wider pores in alginate nanoparticles with minimal cross-linking may mitigate the rapid water release capacity and slow the pace of swelling (Sharma et al., 2014).

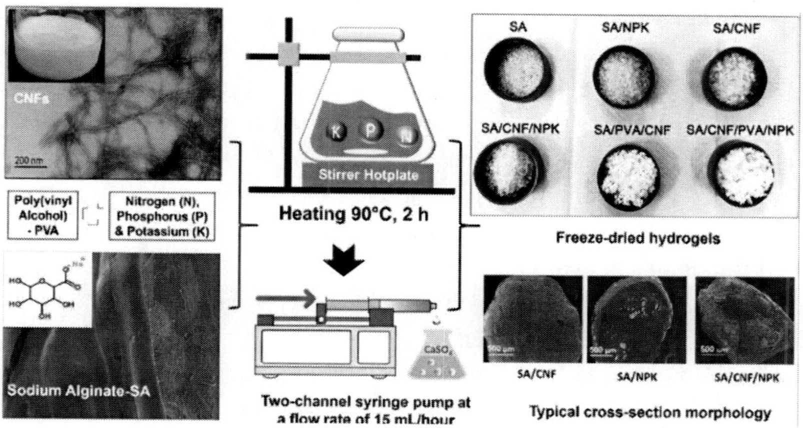

Figure 5. SEM images and schematic illustrations of raw materials, synthesis process, and geometry of the CNF-SA-based hydrogels produced in the 0.1 wt% CaSO4 solution (Adopted from S. Zhang et al., 2022).

Alginate-based composites can improve soil permeability, aeration, and compaction reduction, increase farmland's ability to retain water and lower irrigation frequency. The benefits of alginate-based composites include stable physicochemical features, excellent fertilizer utilization, and long-lasting fertilizer efficiency.

Materials that enhance the physicochemical characteristics and soil structure are generally called soil amendments. However, after fermentation and decomposition, animal excrement, plant and animal wastes, and other byproducts are used as the primary raw ingredients to make organic fertilizer (Jiang et al., 2022).

Most soil amendments are composite compositions with a range of properties and purposes. For example, (Kang et al., 2022) discovered that applying soil amendments based on biochar greatly enhanced the stability of soil aggregates and increased soil water retention by 128.9%, which encouraged the development of maize and beans. Consequently, soil structure can be improved, pH can be regulated, and soil fertility can be raised by applying soil amendments (van der Merwe, Goosen, and Pott, 2022).

Numerous adsorption groups found in alginate-based composites can interact physiochemically in various ways with rhizosphere soil. They can alter the soil's aggregate structure and distribution of pore sizes when applied (Buchmann et al., 2020). By examining the effects of hydrophilic and amphiphilic calcium alginate (SA/Ca and ASA/Ca), respectively, on soil aggregation properties as measured by the aggregate stability index (ADI), rheology, and morphology (Siqi Zhang et al., 2022). validated this viewpoint. The findings demonstrated how well soil aggregation behavior was encouraged by the three-dimensional hydrogel network structure of the SA/Ca and ASA/Ca hydrogels. Furthermore, calcium alginate's amphiphilicity increased its affinity for soil, improving the durability of soil aggregates.

By combining and cross-linking sodium alginate (SA), poly(vinyl alcohol) (PVA), and cellulose nanofibers (CNFs) in the presence of a fertilizer formulation containing nitrogen (N), phosphorus (P), and potassium (K), novel nanocomposite hydrogels were successfully created (S. Liu et al., 2021). The hydrogels had a microporous shell made of semi-interpenetrating polymer networks (IPNs) and a microporous, flexible core. The hydrogel nanocomposite network preserved the NPK compounds' crystalline structure. The SA/CNF/PVA-based hydrogels also demonstrated a greater capacity to retain water in mixed soil and deionized water. Good sensitivity was shown by the swelling behavior in different physiological pH, salt, and alkali solutions. Considering this, the hydrogels produced could be useful for developing precision agriculture and horticulture in the future in regions prone to drought and/or losing their fertilizer supply. Six hydrogels that were based on SA - pure SA, SA/NPK, SA/CNF, SA/CNF/NPK, SA/CNF/PVA, and SA/CNF/PVA/NPK - were created by blending and cross-linking CNFs and SA with and without hydrophilic PVA and NPK fertilizers (Figure 5).

All pure CNF hydrogel samples had swelling ratios that rose quickly in the first few hours (between 0 and 20 hours), mostly because these hydrogels' pore surfaces had a lot of free adsorption sites. At 0.5 weight percent CNF concentration, the maximum swelling ratio was 8.0 g g^{-1} compared to the other four CNF hydrogels. The swelling ratios of the SA/CNF hydrogels increased in the presence of 0.5wt% CNF, dropped as the SA dose increased, and reached their maximum equilibrium ratio of 12.42 g g-1 at 1.5 wt% SA concentration. Samples containing 0.5, 1.0, 2.0, and 2.5 weight percent SA had equilibrium swelling ratios of 95.5, 100.9, 112.4, and 91.6 g g^{-1}, respectively. This finding can be explained by the fact that, whereas lower SA concentrations did not form a three-dimensional hydrophilic network with effective water swelling, higher SA concentrations caused an increased cross-linking density that produced a smaller and closer network space in the polymer matrix to accommodate water swelling. Furthermore, because PVA was more hydrophilic than SA, the equilibrium swelling ratios of PVA/SA hydrogels were generally increased with higher PVA content. More PVA prevented physical cross-linking in these hydrogels containing 1.5 weight percent SA and 0.5 weight percent CNF, which made non-covalent PVA cross-linking more challenging. Consequently, even though all five hydrogels showed comparable tendencies of swelling kinetics, the equilibrium swelling capacity of the SA/CNF/PVA hydrogels reduced with increased PVA concentrations. Three synthesized hydrogels (SA/NPK, SA/CNF/NPK, and SA/CNF/PVA/NPK) were evaluated for their NPK release dynamics in deionized water and mixed soils to better understand the release processes of hydrogel-loaded fertilizers in various media. The SA/NPK hydrogels showed a quick 10-day maximal release in water. The K release in SA/NPK hydrogels was obviously higher than the N and P releases, with the cumulative release ratios of N, P, and K being 0.54%, 0.57%, and 1.00% on the first day and 7.58%, 6.87%, and 8.62% on the tenth. In SA/CNF/PVA/NPK hydrogels, the N, P, and K released 14.03%, 11.95%, and 7.33% in a day and 67.90%, 70.78%, and 71.12% in 30 days. From SA/CNF/PVA/NPK hydrogels, N, P, and K were released at rates of 0.06%, 0.01%, and 0.01% in a day and 24.32%, 36.21%, and 50.17% in 30 days, respectively. The two hydrogels' slow-release characteristics align with the European Committee for

Standardization's slow-release fertilizer standard, as evidenced by the reduction of the target fertilizer's total release by 15% on the first day and up to 75% on the 30th day. All hydrothermal hydrogels' NPK-release patterns in the soil, apart from the SA/NPK and SA/CNF/NPK hydrogels' NPK release, were nearly identical to those in water. This could be explained by the soil's ionic characteristics, which decreased the swelling capacity and, in turn, altered the amount of fertilizer released.

Gelatine

Gelatin holds excellent potential for creating environmentally beneficial biocomposite materials since it is widely accessible, renewable, biodegradable, and biocompatible (Roy and Rhim, 2020). Gelatin has a range of functional groups that include ionizable groups such as the $-NH_2$ and $-COOH$ groups of aspartic acid, the $-NH_2$ group of lysine, the imidazolium group of histidine, the guanidinium group of arginine, and carboxyl and phenolic groups. These groups can be sites for different types of chemical modifications and conjugation opportunities. Because of the large number of functional groups, the properties of GHs and the structure of hydrogel scaffolds can be quickly altered chemically and/or physically by combining copolymers and cross-linkers. The chemical alteration known as acylation by acid anhydrides has been studied in relation to proteins more than any other. Additionally, gelatin has a few unique properties, such as thermo-reversible gel formation (Baseer et al., 2019). One major advantage of gelatin over other polymers is its physical cross-linking capabilities as a hydrogel matrix.

Because of the enormous amount of randomly arranged cross-links in the gelatin chains, pure GHs have limited molecular mobility, making them brittle, fragile, and difficult to utilize in some situations. Thus, techniques for improving GH's properties are being employed; these techniques mostly include combining different materials to create composite hydrogels, which are then reinforced through cross-linking. One significant achievement was the creation of hydrogel nanocomposite materials. Polymeric composite hydrogels, which can be created by mixing several polymers, include simple blend hydrogels, copolymeric hydrogels, interpenetrating networks (IPNs), and double network (DN) hydrogels. In contrast, two or more different types of monomeric units - one of which loves water and causes swelling - compose a copolymeric hydrogel. Because of this, copolymerization techniques are employed to covalently bond the integrated secondary polymers to the hydrogels (both polymers are polymerized or cross-linked), producing fully integrated "hybrid" graphene hydrogels with an array of features. To create reinforced nanocomposite graphene hybrids (GHs), researchers are attempting to incorporate nanomaterials (NMs) into hydrogel systems. Hybrid materials known as nanocomposite GHs are composed of a reinforcing dispersed nanofiller phase (NPs or nanostructures) that are physically distributed or cross-linked with the matrix through physical or chemical means and continuous hydrated 3D gelatin networks with a polymeric phase serving as the matrix. The majority of these nanoscale agents are hydrophilic. The main goals of the research are to increase the necessary features for target applications and to enrich interactions between gelatin chains and NPs. Polymeric hydrogels hold great potential for restoring and reclaiming soil, and their ability to gradually release active chemicals from their matrix makes them an innovative approach to improving agricultural practices and raising agricultural land productivity. Biodegradable carrier-based controlled release fertilizers (CRFs) can lower application rates, stop seeding damage, restrict leaching or volatilization, and protect the environment. Selecting polymers that can function as an appropriate soil conditioner and biodegrade in the soil is crucial when creating CR formulations. The adjustable absorption rate, high water holding capacity, high absorbency, high gel fraction upon cross-linking, cheap cost, non-toxicity, outstanding stability upon swelling and during storage, and re-wettability are all necessary for the perfect hydrogel for agricultural applications. In the field of agriculture, not much research has been done on GHs. The main reason GHs satisfy the criteria is due to the characteristics of gelatin. Due to its high-water absorption capacity (as soil conditioners), GH is primarily used in two categories in the agricultural sector: (1) controlling soil moisture and plant water supply and (2) encapsulating and slowly releasing agrochemicals (ACs) (Mushtaq et al., 2022).

The literature for CRFs reports several GHs. A recent description of the repurposing of gelatin scraps and the creation of gelatin composites incorporating 2-methyl-4-chlorophenoxy acetic acid (MCPA) for use as a herbicide revealed that the final formulation was appropriate for use in agriculture (Kandil et al., 2004).

In a range of pH and saline solutions, the swelling behavior of a super-adsorbent hydrogel based on gelatin-graft-poly (sodium acrylate-co-acrylamide) with salt-responsive properties was investigated (Pourjavadi et al., 2006).

By analyzing the stimuli-sensitive release of agrochemicals, Liu, J., et al., created a pH and Na+ responsive double cross-linked hydrogel membrane of chitosan and gelatin blend to overcome the poor durability and dimensional stability. The films performed well and would be suitable for applications involving separation and water treatment (J. Liu et al., 2020).

The methyl acrylate (MA) and agar-agar copolymerized acrylic acid (AA) hydrogel were investigated as a high-water swelling and soil conditioning material. The hydrogel that was produced demonstrated that the soil's water retention period was extended to 30 days, while the sands were extended to 6–10 days. The absorbent elements are put beneath the soil for a respectable execution. They are frequently used in agriculture for soil bioremediation and as materials to retain moisture (Chaudhary et al., 2020).

Polyacrylamide-based PAAM/gelatin hydrogel converted amide groups ($CONH_2$) to carboxylic groups (COONa) through alkaline hydrolysis with concentrated NaOH. The swelling degree of the chemically altered hydrogels ranged from 220 to 720 ($g\ g^{-1}$). Furthermore, the hydrogel product exhibited a good slow-release (extended to 6 days) capability, as demonstrated by the slow-release capabilities of three different types of fertilizers (K, P, and urea) loaded on modified hydrogels. When applied to irrigation cycles of *Vicia faba* (bean plants) in drought-prone areas, the PAAM-COOK/gelatin soil amendment reduced water stress and enhanced soil water-holding capacity. Hydrogel application, whether on its own or combined with fertilizers, may significantly improve early growth, root establishment, and development. Poplar seedlings treated with hydrogel exhibited longer roots and more surface area (Ghobashy et al., 2018). GHs might help reestablish the arid and desert ecosystem.

(Pulat and Uğurlu 2016) constructed IPN hydrogel systems from various gelatin/polyacrylamide combinations containing maleic (MA) acid to enhance the solubility of phosphate fertilizers. It was found that all hydrogels had swelling values ranging from 435 to 830 percent. The release of MA from loaded hydrogels demonstrated that mixing GHs loaded with MA effectively reduced pH and increased the solubility of $Ca_3(PO_4)_2$ in the releasing medium.

The effects of gelatin on the properties of hybrid hydrogels were investigated by (Rattanamanee et al., 2013). The percentage nitrogen release of silk fibroin and SF-gelatin hybrid hydrogels with mass ratios of 25:75, 50:50, and 75:25 was 46, 80, 59, and 51%, respectively. This indicates that the hydrogel's gelatin and SF composition affects the amount of nitrogen released. A decrease in SF crystallinity, associated with the hydrogel's swelling behavior, was seen when the quantity of gelatin in the blended hydrogel was increased. Moreover, the SF-gelatin hydrogel has a lower release rate (k) and diffusion coefficient (D) than pure SF-based gel, suggesting that it could be utilized for nitrogen capture. Additionally, this study team reported on the regulated release of urea by applying a ternary blend consisting of SF, gelatin, and chitosan.

They also showed that increasing the CS content in the hydrogel of SF/gelatin blend decreased porosity, affecting the degree of swelling, crystallinity, surface hydrophobicity, and water solubility. The amount of CS present had a similar effect on the rate of urea release from hydrogel; for ternary blended hydrogels, the Korsmeyer-Peppas model yielded a diffusion exponent characteristics (n) value bigger than 1.0. Interestingly, the hydrogel's fertilizer release rate (k) was lower than that of the previously published SF/gelatin hydrogel alone, suggesting that CS may be applied to prolong the release of urea from the hydrogel for applications requiring sustained release. Chitosan was utilized to alter the SF/gelatin blend hydrogel's network structure to increase the release rate (Rattanamanee et al., 2015).

Synthetic SAPs

In general, SAPs composed of synthetic polymers - also known as super-adsorbent resin (SAR) - like PAA and PAM have better water-absorption and water-retention qualities than SAPs comprised of natural polymers. Nevertheless, synthetic polymers cannot break down in soil after many years and can seriously contaminate it. Thus, the opposing relationships between absorbency, water retention and biodegradability can be balanced by blending synthetic SAPs with natural polymers.

Polyacrylic Acid (PAA)

Due to their greater water absorbency (300-500 g g^{-1} in water) and water retention capacity, polyacrylic acid (PAA) and sodium polyacrylate SAPs are the most utilized water retention agents. However, their non-degradability, poor mechanical strength, and low salt resistance limit their efficacy. It has recently been discovered that adding clay, such as kaolin, montmorillonite, and bentonite, can improve the mechanical performances of SAPs and increase their water absorbency. As an illustration, a slow-release fertilizer with a semi-IPN structure based on linear urea-formaldehyde oligomers (UF) and kaolin-g-P(AA-co-AM) demonstrated excellent water absorbency of 68 g g^{-1} in tap water and controlled fertilizer release (Xiang et al., 2017). In distilled water, a hybrid linear low-density polyethylene (LLDPE)-g-PAA/montmorillonite combination demonstrated an optimal water absorbency of 800 g g^{-1} (Irani, Ismail, and Ahmad 2013). A semi-IPN structure with enhanced mechanical stability and optimized water absorbency of 1018 g g^{-1} in distilled water and 71 g g^{-1} in 0.9 wt% NaCl solution comprised a network of cotton stalk (CS)-g-PAA/bentonite (bent) and linear polyvinylpyrrolidone (PVP) and urea (Wen et al., 2016). Extensive research has been conducted on incorporating natural polymers into the cross-linked copolymer systems of acrylamide (AA) and 2-acrylamido-2-methyl propyl sulfonic acid (AMPS) to attain higher degradability and absorbency. For instance, P(AA-co-AMPS) was grafted with sodium humate (SH), a naturally occurring organic substance that contains enormous amounts of carboxylates and phenolic hydroxyls. In comparison to plain P(AA-co-AMPS) of 600 g g^{-1} in distilled water and 60 g g^{-1} in 0.9 wt% NaCl solution, the generating P(AA-co-AMPS)/SH showed noticeably higher water absorbency of 1380 g g^{-1} in distilled water and 117 g g^{-1} in 0.9 wt% NaCl solution (Fekete et al., 2017). In addition, it was shown that SH added to the surface of P (AA-co-AMPS) might boost water absorbency (1000 g g^{-1} in distilled water and 80 g g^{-1} in 0.9 wt% NaCl solution) (Guan et al., 2018). Adding chitosan to P(AA-co-AMPS)/SH resulted in an optimum water absorbency of 1097 g g^{-1} and 103 g g^{-1} in deionized water and 0.9 wt% NaCl solution, respectively (Yu et al., 2019). This was due to the modest amount of chitosan added to the water absorption capacity. According to the degradation performance test, after 100 days, the product's rate of deterioration in soil may reach 31.9%. The hybrid product exhibited optimal water absorption of 1329 g g^{-1}, 269 g g^{-1}, and 140 g g^{-1} in distilled water, tap water, and 0.9 wt% NaCl solution, respectively, upon the addition of hydroxymethyl cellulose sodium (NaHMC) and laterite to P(AA-co-AMPS) (Cheng et al., 2019). The HPC-g-P(AA-co-AMPS)/laterite that was produced after substituting hydroxypropyl cellulose (HPC) for NaHMC exhibited water absorption of 1427 g g^{-1}, 272 g g^{-1}, and 143 g g^{-1} in distilled water, tap water, and 0.9 wt% NaCl solution, respectively. Additionally, the water absorption was 946 g g^{-1} after five repetitions of absorption, suggesting promising application potential (Cheng et al., 2020).

Water-soluble toxic elements that were still present in SAPs could potentially endanger crop growth by lowering yield, inhibiting root system growth, and decreasing seed germination rate. These limitations prevent it from being widely used in forestry and agriculture. In this work, modified diatomaceous earth was added to increase the grafting efficiency and ammonia solution was used in place of strong alkali sodium hydroxide for neutralization, resulting in the synthesis of a novel class of super absorbent polymers (AD-SAP) with minimal residual acrylic acid and sodium ions. The reaction mechanism of AD-SAP is shown in Figure 6. The sodium ion content in AD-SAP was reduced to 0.12 g kg^{-1}, which is significantly lower than the three commercially available SAPs, and the residual acrylic acid content was 0.03% when 28% of acrylic acid, 24% of ammonia solution, and 5% of diatomaceous earth were employed for synthesis. Its ability to absorb water has neither diminished nor its resistance to salt has strengthened. AD-SAP has a 578 g g^{-1} water absorption ratio and an 87 g g^{-1} salt absorption ratio. Pot experiments demonstrated that AD-SAP removed the detrimental effects of commercial SAPs on crop development. The study offered scientific advice on how to lessen the detrimental effects of SAPs on crop growth and provide technological support for the manufacture of SAPs (Situ et al., 2023).

This study primarily intends to develop a form of SAP with minimal residual acrylic acid and Na^+ and to provide insights for optimizing the SAP production process in agriculture and forestry. The goal is to prevent crop damage caused by the hazardous components that remain in SAPs after production. An ammonia solution was selected to neutralize the acrylic acid and lower the amount of Na^+ and K^+ cations in SAPs. Diatomaceous earth was utilized as an inorganic element to create the organic-inorganic composite SAP. This increased the grafting efficiency and decreased the amount of leftover acrylic acid. The effects of three commercially available SAPs and ammonia solution-diatomaceous earth composite SAP (AD-SAP) on maize development under no-water-stressed settings (normal conditions without water stress) were also compared in a pot experiment.

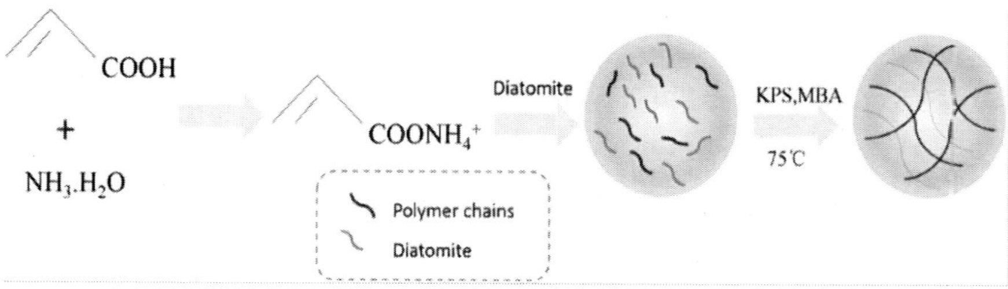

Figure 6. The reaction mechanism of AD-SAP (Courtesy of Situ et al., 2023).

The water-retaining agents included water-soluble hazardous chemicals in rather high concentrations; the residual acrylic acid ranged from 0.14% to 0.72%, and the residual Na^+ from 10.6 g kg^{-1} to 12.6 g kg^{-1}. Acrylic acid exhibited inhibitory effects on crops at concentrations as low as 2.5 mg l^{-1}, primarily suppressing root growth. According to previous research that included water culture experiments and seed germination tests, the inhibitory effects were amplified in the presence of both acrylic acid and Na^+. A pot experiment conducted in an environment without water stress was used to assess the impact of the water-soluble components in the four varieties of SAP on crop growth. Maize treated with acrylamide-acrylic acid SAP, potassium acrylate SAP, and sodium acrylate SAP showed a considerable reduction in stem and leaf biomass. They dropped 12.6% to 41.5%, and the root biomass dropped 30.3% to 46.7% compared to the control (CK0). This illustrates how SAP may hinder crop growth. According to the biomass, which aligns with other research findings, sodium polyacrylate SAP (CK1), one of the commercially available SAPs, harms crops the most. This order of biomass is CK2 > CK3 > CK1. Nevertheless, there was no discernible variation in biomass between the AD-SAP (T1) and CK0 treatments. The trends in biomass at 15 and 30 days were comparable. The biomass of CK1, CK2, and CK3 decreased, although T1 displayed growth-promoting characteristics. This could be because the AD-SAP contains ammonia and diatomaceous earth, both of which stimulate crop development in a certain way. This study confirms that the remaining water-soluble components in SAP are one of the causes of the unstable application effects of SAP in agriculture by showing that lowering the concentration of acrylic acid and Na^+ in SAP can mitigate the detrimental impacts on crops. Research is done on how various SAP types affect the morphology of maize roots. The maize root growth was decreased, and the increase in root surface area was suppressed by the CK1, CK2, and CK3 treatments compared to CK0. A notable suppression of root development was also observed. Regarding the length and surface area of the maize roots, there was no discernible difference between the T1 treatment with AD-SAP and CK0. This suggests that the shape of maize roots was only slightly inhibited by the administration of AD-SAP. The primary manifestation of acrylic acid's harmful effects on crops is its suppression of root growth and corrosive effects on the roots. While treating AD-SAP with low residual levels of acrylic acid and Na^+ did not demonstrate inhibition, applying the three types of commercial SAPs inhibited the root length and surface area of maize roots. This validates the detrimental effects of acrylic acid on crop growth. On the mineral components in maize plants' stems, leaves, and roots, the effects of potassium polyacrylate, sodium polyacrylate, and acrylamide-acrylic acid SAP were largely identical. The Ca^{2+} and Mg^{2+} levels dropped, the Na^+ levels rose, and the K^+ levels barely changed. K^+ levels, however, increased with the potassium polyacrylate SAP (CK2) treatment and decreased with the sodium polyacrylate SAP (CK1) and acrylamide-acrylic acid sodium SAP (CK3) treatments. Commercially available SAPs impact the availability of exchangeable nutrients in the soil. This includes a decrease in exchangeable Ca^{2+} and Mg^{2+}, an increase in Na^+, and a decrease in the absorption of Ca^{2+} and Mg^{2+} content due to damage to plant root systems. There were no appreciable variations in the levels of Na^+ and K^+ as compared to CK0 in the AD-SAP (T1) treatment, which did not include any addition of these elements in the material. Due to the presence of some Ca^{2+} in the diatomaceous earth, which causes greater Ca^{2+} accumulation in the roots, the accumulation of Ca^{2+} in the roots after 30 days was 76.0% to 105.1% higher than in CK0. The application of acrylamide-acrylic acid SAP, potassium polyacrylate SAP, and sodium polyacrylate SAP prevented the plants from accumulating N and P. At 30 days, N uptake in the stems and leaves was lower in the CK1, CK2, and CK3 treatments than in the CK0 treatment - decreases of 34.6%, 30.1%, and 48.0%, respectively. Similarly, compared to CK0, the P uptake in the stems and leaves dropped by 43.7%, 14.5%, and 36.1%. Compared to CK0, N uptake in the roots dropped by 40.7%, 45.8%, and 55.2%, and P uptake dropped by 58.7%, 51.8%, and 61.4%. As opposed to CK0, the T1 treatment, which included AD-SAP, demonstrated a marginally higher rate of N uptake in the stems and leaves, ranging from 0.7% to 9.1%; however, this difference was not statistically significant. Compared to CK0, the T1 treatment resulted in a 33.3% to 39.5% increase in P absorption in the stems and leaves. While there was no discernible difference in the P uptake between the roots and CK0, the N uptake was marginally lower. Root development restriction inhibits the uptake of N and P, two important nutrients for plant growth. Because AD-SAP has a higher N concentration, it facilitates P absorption.

A novel water-harvesting hydrogel composite reinforced with bentonite clay and modified cellulosic/acrylic acid has been created. Because of their exceptional super absorbency and high retention capacity, these unique super-adsorbent hydrogels have been investigated for their potential to support plant growth and survival, especially in semi-arid regions. When the hydrogel with 5% clay reaches equilibrium swelling after a week, it shows 36600% absorbency and a maximum swelling percentage of 19040% within 26 minutes at 25°C. Brassica juncea seedlings are grown in a mixture of this super-adsorbent hydrogel and black soil that is cellulosic and biocompatible. This report also includes a detailed three-week observation with a modified planned watering schedule replicating the semi-arid region's drought conditions. Even during the period of decreased watering for the gel-soil mixture, a notable degree of plant survival and growth is seen (Chaudhuri et al., 2022).

By solution polymerizing acrylic acid in the presence of carbamoyl ethyl ether of cellulose (CEEC) and varying proportions of bentonite clay with 1 weight percent cross-linker (MBA) and 1 weight percent initiator (KPS) concerning total monomer content, composites based on acrylic acid (AA), CEEC, and Bentonite clay were created. Bentonite was used in the preparation of the composites, which were given the following designations: CAA0 (0 weight percent clay), CAA1 (5 weight percent clay), CAA2 (10 weight percent clay), CAA3 (15 weight percent clay), and CAA4 (20 weight percent clay). After being submerged in distilled water at room temperature for 26 minutes, it was noticed that the hydrogels based on CEEC had attained their equilibrium condition. However, CAA1 swells to its maximum (190.4 g g^{-1}) in 26 minutes. It has been discovered that CAA1 has the highest equilibrium swelling ratio (E.S.R.) (185.01 g g^{-1}). Concerning the CAA0, which contains 0 weight percent clay, it is anticipated that there may be more entanglements or physical cross-links between the clay particles and the corresponding hydrogel under consideration in the composites as the proportion of bentonite increases from 0 to 5 weight percent and then to 15 weight percent. Because of these physical cross-links, the amount of bentonite clay in the amorphous polymer matrix may have increased, leading to an increase in stiffness at low temperatures. The degree of chain rigidity, which is assumed to be increased by a higher proportion of clay incorporation, can be linked to swelling efficiency.

Reusing agricultural waste is a crucial tactic for ensuring contemporary society's sustainable growth. In this work, a novel slow-release fertilizer of corn stalk fiber/geopolymer (N @ CoSP/LoGP) was obtained by combining corn stalk fiber (CoS), a typical biomass that is discarded from agricultural production, with loess-based geopolymer (LoGP), loading urea, and *in-situ* grafting polymer of acrylic acid. The maximal swelling rates of N @ CoSP/LoGP in water and 0.9 weight percent NaCl solution were 263.2 g g^{-1} and 62.5 g g^{-1}, respectively, indicating that the created composited fertilizer might improve the soil's ability to retain water. After six cycles, its swelling rate remained at 80% of the initial value, and the soil's water-holding capacity reached more than 25 days despite having 2% of N at CoSP/LoGP. Additionally, pot studies with maize were used to investigate N @ CoSP/LoGP's ability to promote plant development. The created composited fertilizer released 60.6% of its nutrients within 45 days, indicating that it may successfully encourage the germination and growth of maize crops (H. Yang et al., 2023). Figure 7 depicts the procedure for preparing N @ CoSP/LoGP. To bond the LoPs particles by Si-O-O-Si/Al, the aluminosilicate on their surface was first activated using a silicate alkali solution (Na2SiO3-NaOH) as an alkali activator. Sesame oil was utilized as an alkali activator because a portion of its glycerin fatty acid ester was hydrolyzed in an alkaline solution. Furthermore, during the drying process, the fatty acid sodium emulsion produced was favorable to the successful isolation of LoPs particles. It was further immersed in warm water to remove the salt, a higher fatty acid, and create pores in the resulting LoGP after it had dried and molded. Second, loading urea and LoGP were combined with CoS fiber. Additionally, AA underwent *in-situ* polymerization while being cross-linked by MBA. In addition to preventing dense polymer formation and the self-polymerization of highly active AA, sodium hydroxide neutralization of AA also raised the polymer's osmotic pressure and hydrophilic group (–COONa), both of which accelerated the polymer's swelling rate.

It was discovered that N @ CoSP/LoGP may expand slowly in 0.9% NaCl solution but quickly in distilled water after being submerged for five minutes. The study concluded that the swelling behavior of N @ CoSP/LoGP in water was superior to that of saline solution. It showed that in stronger acid (pH: 2) and alkaline (pH: 12) media, the swelling ratio of N @ CoSP/LoGP was smaller than in other pH values. On the other hand, at a pH of 6, the swelling ratio of N @ CoSP/LoGP peaked. It showed that N @ CoSP/LoGP had a higher swelling rate and equilibrium water absorbency in the low-concentration NaCl solution. The osmotic pressure of the internal and exterior environments in N @ CoSP/LoGP dramatically decreased as the concentration of Na^+ in the solution increased, which in turn affected the water absorbency performance. In the meantime, thermal movement made it simpler for the high concentration of Na^+ to enter the substance and form an electrostatic bond with the hydrophilic group. It caused the internal hydrophilic group's hydrophilic action to diminish, which in turn diminished the swelling impact. Early in the release process, it was noted that urea had attained full release. Nearly 90% of the nitrogen in commercial fertilizers and urea was discharged within ten days. During the course of 15 days, the nitrogen of N @ CoS-PAA release reached equilibrium, and the cumulative released rate increased to 64.1%. In contrast, N @ CoSP/LoGP demonstrated a superb sustained-release capability; 60.6% of the nutrients were released in 45 days, a 9.2% increase over earlier trials.

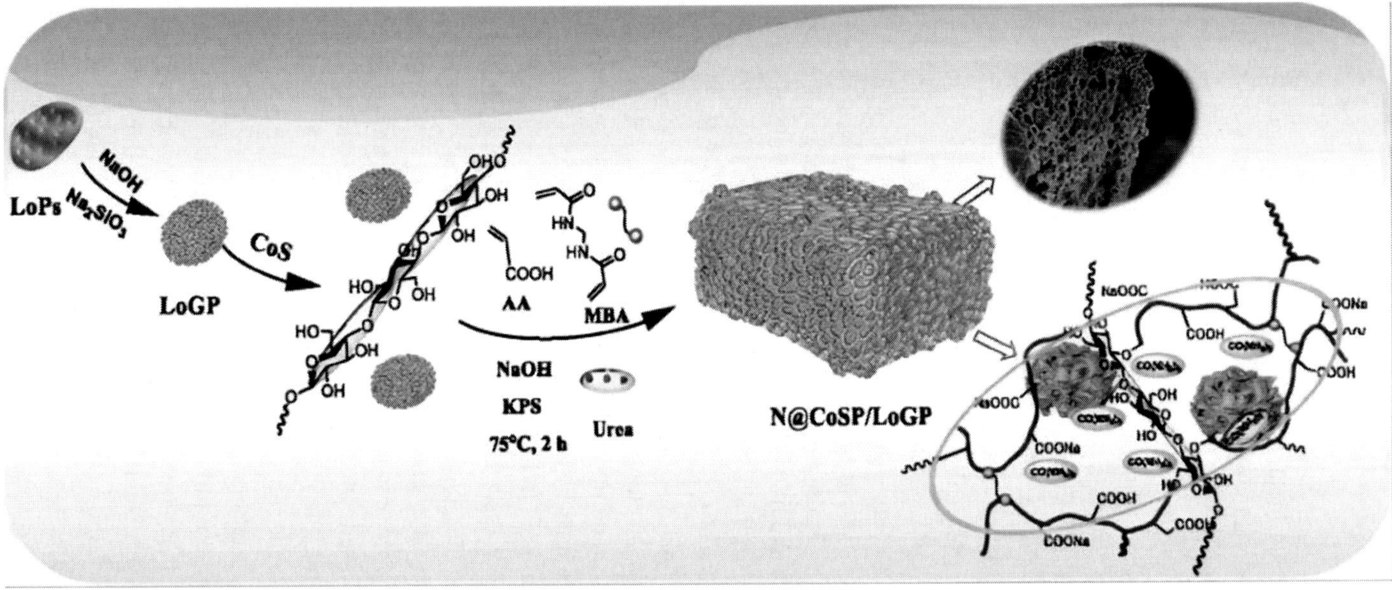

Figure 7. Procedure for synthesis of N @ CoSP/LoGP (Courtesy of H. Yang et al., 2023).

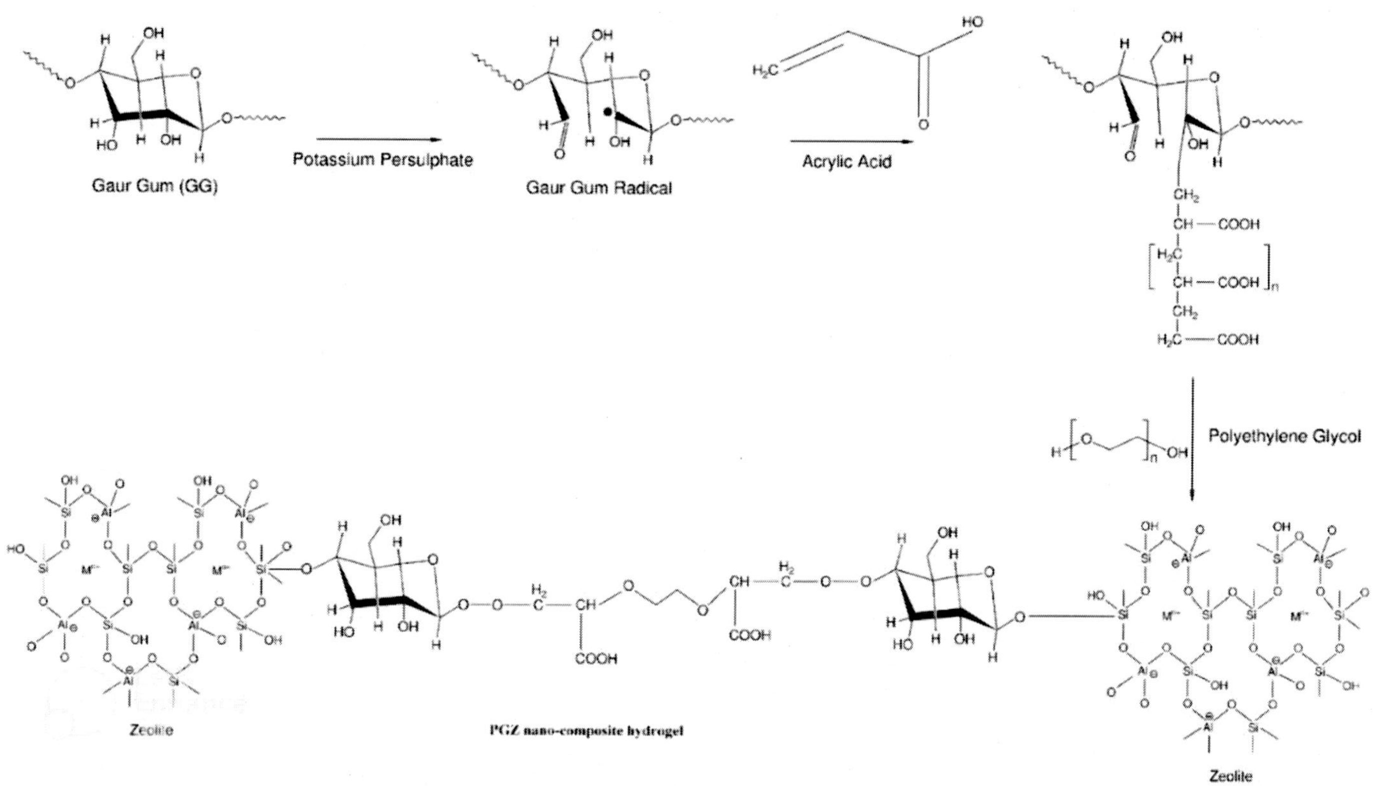

Figure 8. Synthesis mechanism of PGZ nano-composite hydrogel (Courtesy of Songara, Patel, and Mungray 2022).

Another study used zeolite and guar gum (Zeolite-GG) to create a nano-composite hydrogel based on natural ingredients. The cross-linker used in the development of the PAA/GG-Zeolite nano-composite hydrogel (PGZ nano-composite hydrogel) is polyethylene glycol (PEG-600) combined with acrylic acid (AA). The synthesis mechanism of PGZ nano-composite hydrogel is demonstrated in Figure 8. It was discovered that the produced hydrogel has a swelling index of 92.50 mL per gram. Because of its greater potassium (K) content and high-water absorption ability, the synthesized PGZ Nano-composite hydrogel has enough potential to be used in agriculture as a nutrient carrier and soil conditioning agent, especially in dry and semi-arid regions (Songara, Patel, and Mungray, 2022).

When a PGZ nano-composite hydrogel was applied, the soil profile's moisture content was higher than it would have been otherwise. Following the initial irrigation, the soil moisture content measured by applying PGZ nano-composite hydrogel was 18% higher than without hydrogel (control). The PGZ nano-composite hydrogel had preserved the 70% humidity in the vicinity of the rose pot. A comparison between commercially available hydrogel and PGZ nano-composite hydrogel was also conducted in terms of monitoring moisture. When the rose received its first watering at the initial level, the soil moisture measurement was 100%. The moisture level will then gradually decline after that. Thus, before applying a second watering to the rose pot, the moisture content preserved by commercially available hydrogel was 25%, that of PGZ nano-composite hydrogel was 28%, and that of the rose pot without hydrogel was just 10%. The rose pot receives its second watering after eight days, and after a few days at a time, the moisture level is once more tested. After nine days, the final moisture content readings were 20%, 25%, and 10%, respectively. It shows that soil moisture declines quickly during both watering when the hydrogel is not applied.

Graft copolymerization of acrylic acid (AA) and 2-hydroxyethyl methacrylate (HEMA) onto sodium alginate in the presence of montmorillonite (MMT) produced Na-alginate-g-poly (acrylic acid-co-2-hydroxyethyl methacrylate)/montmorillonite super-adsorbent composites (scheme 9). According to the findings, the composites' maximum swelling capacity in distilled water was 752 g g^{-1} when 1.25% KPS and 0.06% MBA, 75% AA, 6.25% HEMA, and 12.5% Na-Alg were used. To assess the release potentials of various samples, urea was added, and the conductivity of the samples was measured to examine the release. As the amount of MMT grew, so did the amount of urea emission (Kenawy, Azaam, and El-nshar, 2019).

The findings demonstrated that a 16-day quick release was followed by a continuous release. The release of urea that was loaded close to the surface of the superabsorbent composite (SAC), which its preparation process is demonstrated in Figure 9, may have caused the initial rapid release. The releasing behavior in an acidic and basic medium differs slightly.

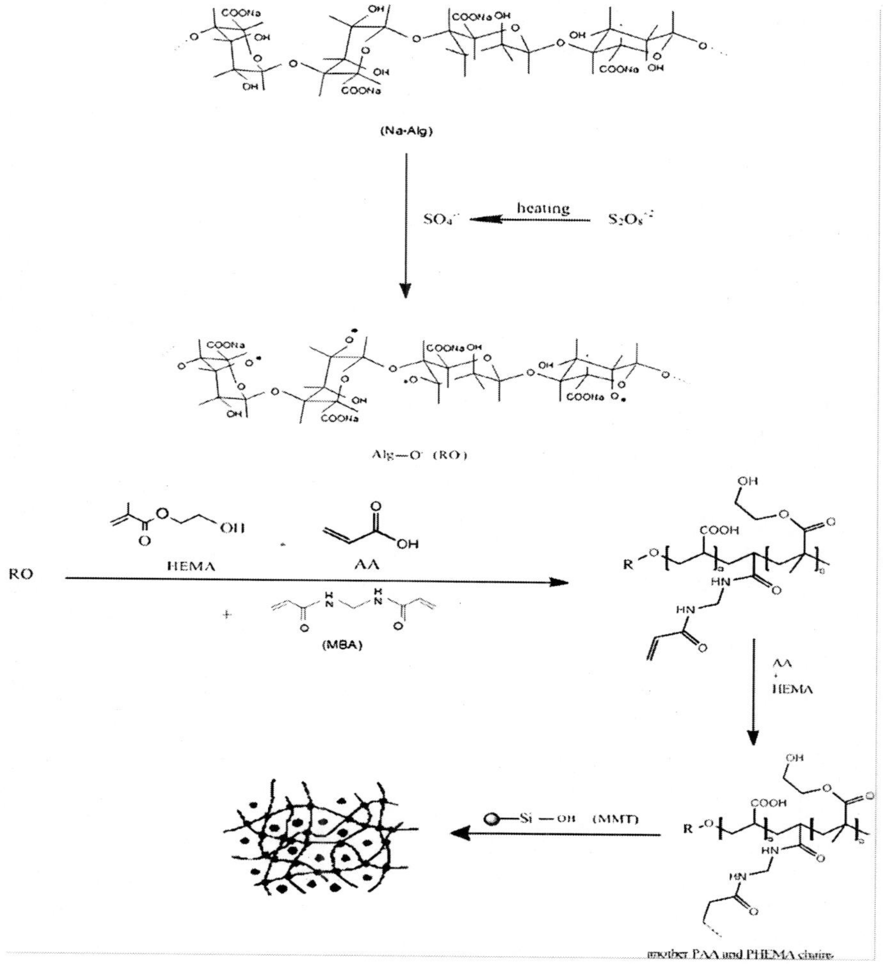

Figure 9. Preparation of super-adsorbent composites (SACs) (Courtesy of Kenawy, Azaam, and El-nshar, 2019).

Polyacrylamide (PAM)

After PAA, polyacrylamide (PAM) is the second most widely used synthetic SAP. It absorbs water similarly to PAA. Its application is, however, severely limited by its weak mechanical qualities and low salt tolerance. As a result, numerous tactics have been put out to improve PAM's overall effectiveness. After creating poly(AM-co-vinyl alcohol)/clay composites, researchers investigated the effects of several clay types on poly(AM-co-vinyl alcohol) copolymeric super-adsorbent materials. Adding 2 weight percent Cloisite®30B clay produced the best water absorbency ratio of 5455%. Despite having a lesser water absorbency than pure PAM, it demonstrated strong resilience to acids and alkalis. After a 24-hour period, the swelling was 46% less in 1 M HCl and 37% less in 1 M NaOH as compared to double-distilled water. Additionally, this material demonstrated positive re-swelling, allowing it to absorb more water than before (Swain, Shur, and Patra 2013). A starch-g-PAM super-adsorbent was created to boost PAM's water absorbency; it demonstrated maximum absorption of 1452 g g^{-1} and 83 g g^{-1} for distilled water and 0.9 wt% NaCl solution, respectively (Sufen Zhang et al., 2014). Based on this, the soluble starch-g-P(AM-co-NaAMC14S) that was produced upon grafting the hydrophobic monomer sodium 2-acrylamido-tetradecane sulfonate (NaAMC14S) to the copolymerization of soluble starch and AM showed the maximum water absorbency of 2537 g g^{-1} in distilled water and 93 g g^{-1} in 0.9 wt% NaCl solution. This suggests that NaAMC14S has a great deal of potential to increase the water absorbency of super-adsorbent resin. Furthermore, because of hydrophobic interaction, the mechanical characteristics of SR-14 have significantly enhanced since the addition of NaAMC14S. This study also discovered that the water absorbency of the AM system can be improved by adding NaOH-induced hydrolysis to the final stage of gel polymerization (Lan et al., 2019).

Graft copolymerization of acrylamide (AM) onto sugarcane bagasse (SCB) in the presence of attapulgite (ATP) produced sugarcane bagasse-g-poly(acrylamide)/attapulgite super-adsorbent composites (SBCs) (Figure 10). Investigations were conducted into the impact of the initiator, cross-linker concentrations, and attapulgite/hydrogel weight ratio (0.0–14.4 wt%) on the swelling capacity. The process of creating sugarcane bagasse (SCB)-g-poly(acrylamide)/attapulgite super-adsorbent composites (SBCs) involved grafting acrylamide (AM) onto SCB while attapulgite (ATP) was present (Figure 10). On the swelling capacity, the impact of the initiator, cross-linker concentrations, and attapulgite/hydrogel weight ratio (0.0–14.4 wt%) was examined (Kenawy et al., 2020).

A super-adsorbent composite (SAC) was created using carboxymethyl cellulose, acrylamide, and montmorillonite in a free-radical graft copolymerization process (Figure 11). the elements impacting the composite's swelling capacity were identified to achieve the high swelling capacity. Due to its high swelling capacity, the composition (15 weight percent carboxymethyl cellulose, 5.4 weight percent montmorillonite, 82 weight percent acrylamide, 0.07 weight percent N,N-methylenebisacrylamide, and 1.1 weight percent potassium persulfate) was chosen to be loaded with urea fertilizer, and the release was examined by measuring the conductivity. According to the results, the new controlled-release method offers good slow-release properties (Kenawy, Azaam, and El-nshar, 2018). As the amount of MMT increased, the emission of urea reduced. For 0 weight percent MMT clay, the total nitrogen release from CMC-g-PAM was 96% in 83 days. Subsequently, the release from CMC-g-PAM/MMT [2.7, 5, 10, 14.4 weight percent MMT] was 86.3, 78.8, 74.28, and 70.2%, in that order. These data confirmed the good, controlled release feature of the CMC-g-PAM/MMT compound. The findings demonstrated a rapid release for the first sixteen days, followed by a continuous release. The loaded urea release close to the SAC surface was the cause of the first rapid release. The releasing behavior in an acidic and basic medium differs slightly.

Using rice husk (RH) as a filler, gelatin, acrylamide (AM), and acrylic acid (AA) were free radical copolymerized to create low-cost composites with good water absorption capacity (Kenawy et al., 2021). Investigations were conducted into the effects of many aspects, including the composition of the composites and their water absorption capacity, the alkaline treatment of the rice husk, and the salinity and pH of the media. Tests on soil biodegradation and water retention were also conducted. The optimized composite with treated RH exhibited good resistance in saline solutions and the pH range of 6–10, with a maximum water uptake value of 825 gg_1 in distilled water. After 30 days, samples of soil containing 1 weight percent of super-adsorbent composites still held almost 7% of the water that had been initially absorbed, indicating positive impacts on soil water retention. A 60 kgha_1 dosage of SAC enhanced the growth characteristics, photosynthetic pigments, and nodulation properties of drought-stricken peanuts. Based on the favorable outcomes of water retention and regulated biodegradability in the soil, the created composites hold promise for improved water management in agriculture and associated environmental advantages. In contrast, the soil with

0.25, 0.5, 0.75, and 1 g of super-adsorbent per 100 g of soil lost all its absorbed water after 17, 19, 22.5, and 32 days, respectively. The findings of the experiment demonstrated that the soil lacking super-adsorbent lost all its absorbed water after 14 days. As predicted, the rate of water loss decreases with increasing SAC/soil dosage. The sample showed a consistent weight reduction over time, and after 160 days, or roughly five months, a total weight loss of almost 60% was attained, demonstrating the SAC prepared in this work's good degradability in soil. Water-stressed peanut growth parameters were effectively improved by soil amendment with different doses of SAC. The stressed peanut's growth rate improved when the SAC dose was increased from 15 to 60 kgha_1, after which it slightly fell with the highest dosage (90 kgha_1). Growth was dependent on the SAC dosage up to 60 kgha_1.

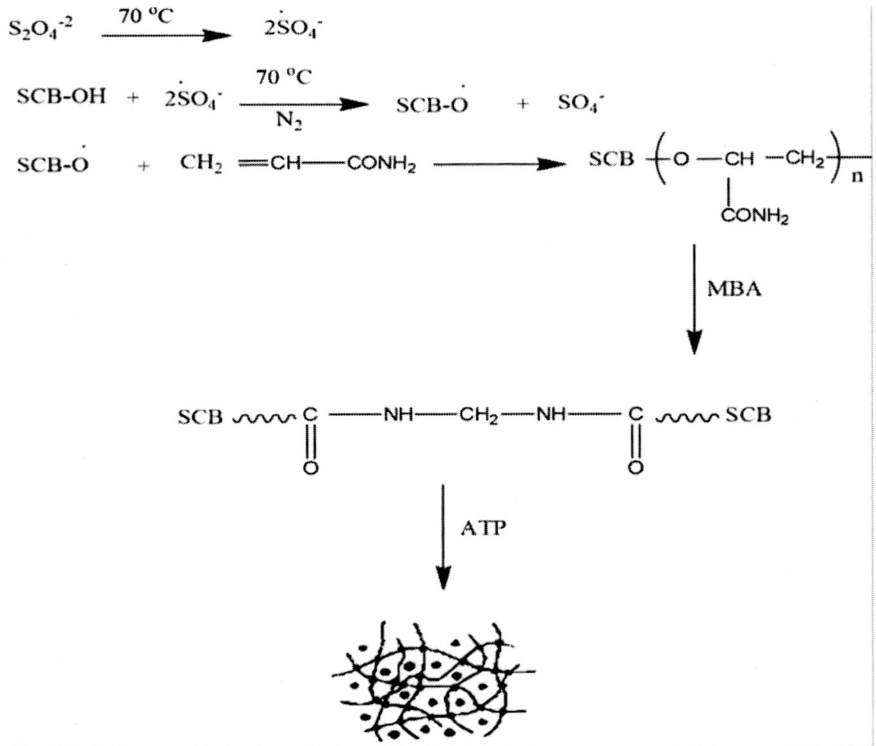

Figure 10. The suggested grafting mechanism for AM and SCB (Courtesy of Kenawy et al., 2020).

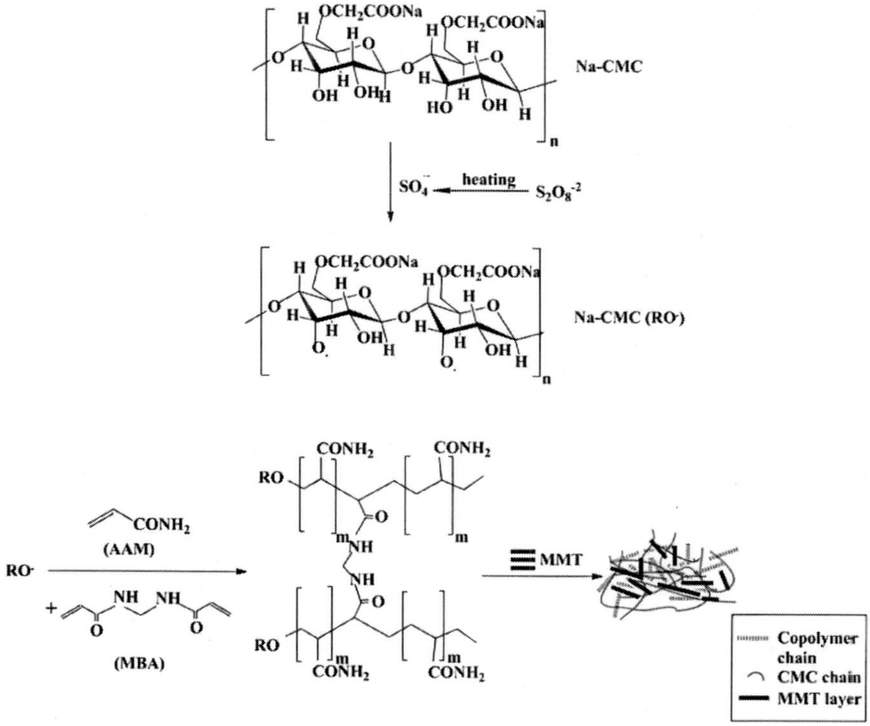

Figure 11. Proposed mechanism for graft copolymerization of acrylamide, carboxymethyl cellulose, and montmorillonite (Courtesy of Kenawy, Azaam, and El-nshar, 2018).

Through the cross-linking of N,N'-methylene-bis-acrylamide (MBA) with chitosan (Ch), acrylamide, and attapulgite (ATP), super-adsorbent composites (CTS) consisting of g-poly(acrylamide)/attapulgite were produced (Kenawy, Elnaby, and Azaam 2024). The study aimed to ascertain the effects of several parameters on the composite's swelling capacity, including the initiator %, cross-linker ratio, and clay content. The outcomes demonstrated that adding ATP increased the composite's thermal stability. When the KPS concentration reached 0.1 g, the maximal swelling was reached. On the other hand, 0.2 g of KPS added to the composite resulted in a reduced swelling capacity. The swelling increased from 210 to 319 g g^{-1} when the amount of ATP was increased by up to 0.4 g. Nevertheless, the swelling capacity dropped to 170 g g^{-1} as the clay amount was raised to 1.2 g. The swelling of ATP2 increased when pH rose to 3.0 but decreased as pH varied from 3 to 6. The pH increased significantly to a pH of 8. The selected composition swelled to 319 g g^{-1} in water (CTS2) and was assessed as a controlled release method for urea fertilizer. The results showed that the release rate was delayed, and the release duration increased from 5 to 21 hours when the formulation's ATP concentration was raised from 0 to 1.2 g.

The Copolymer of Acrylic Acid and Acrylamide (P(AA-co-AM))

The copolymer of acrylic acid with a tiny quantity of acrylamide (P(AA-co-AM)), which is frequently called "SAP" in the literature, is typically the most prevalent SAP on the market (Ahmed Khan et al., 2020). P (AA-co-AM) has been widely used in various industries, including agriculture, commodities, medicine, and health, due to its excellent water absorbency, big water retention capacity, and ease of accessibility. The water absorption of sulphoaluminate planting concrete was 74% higher than that of the reference at the addition level of 1.5 wt% P(AA-co-AM), and the water release rate was 26% lower at 180 hours (L. Li et al., 2017). However, to use less degradable P(AA-co-AM), its water absorbency needs to be further increased. Many studies focus on compounding inorganic particles with P(AA-co-AM) to accomplish this. For instance, photocatalytic surface-initiated polymerization was used to create TiO_2-P(AM-AA), which successfully addressed issues including poor interfacial combination of organic and inorganic and uneven dispersion of inorganic particles in the organic phase. The mechanism of this technique was revealed by IR and XPS detection. Under UV irradiation, TiO_2 created free radicals on its surface, which encouraged an *in-situ* copolymerization reaction between AM and AA monomers on the surface of TiO_2. After the cross-linking agent was added, a three-dimensional network was created. According to the results, the product's maximum water absorbency is 70 g g^{-1} in a 0.9 weight percent NaCl solution and 942 g g^{-1} in deionized water (Gao et al., 2021). Additionally, $Ca(OH)_2$ nanoparticles can increase the AA-AM copolymer's swelling ratio. After three days in the open at 298 K with a humidity of about 40%, the maximum swelling ratio of the resultant PAM/PAA/$Ca(OH)_2$ reaches high up to 5580 g g^{-1} at pH = 7 and 328 K. The water retention ratio also stays over 60%. Additionally, this SAH sample can also be made to have an excellent water-absorbing ability in 0.9 weight percent NaCl solution and 0.5–20 weight percent urea solution (Lv, Wu, and Shen 2019). This is anticipated to have applications in various industries, including wastewater treatment, agriculture, and sanitary products. Notably, the amount of soil present can drastically lower SAPs' ability to retain water. After one hour, this value amounted to 5 g g^{-1} for a 30 cm deep layer of soil with a bulk density of 1.3 g cm^{-3}, and more than 200 g g^{-1} for the control sample that was not loaded. Additionally, loads increase P(AA-co-AM)'s swelling time. For a load of 3.8 kPa, the rate parameter—the amount of time needed to reach 63% of maximal absorption capacity—rose from 63 minutes for the control sample to more than 300 minutes (Lejcuś, Śpitalniak, and Dabrowska 2018). The glycerol-lactic acid-based star-shaped modifier (SM) was used to modify the surface of P (AA-co-AM) due to its poor saline-absorbency under load and worse swollen gel strength. According to the results, surface treatment raised the saline absorbency under load by 28% because of a greater elastic modulus. However, it lowered the water absorbency in both deionized water and saline solution without load because of a higher surface cross-linking density (Ghasri et al., 2019).

Conclusion

Super-adsorbent polymers are gaining a lot of attention as controlled-release, slow-release fertilizers for agricultural applications due to their enhanced water absorption and retention characteristics. This chapter covers several ways to prepare natural and synthetic super-adsorbent polymer composites and how they are used in the realm of agriculture.

Disclaimer

None.

References

Ahmed Khan, T., Munirah Ezzah Tuan Zakaria, Hyun Joong Kim, Suriati Ghazali, and Saidatul Shima Jamari. 2020. "Carbonaceous Microsphere-Based Super-adsorbent Polymer as Filler for Coating of NPK Fertilizer: Fabrication, Properties, Swelling, and Nitrogen Release Characteristics." *Journal of Applied Polymer Science* 137 (8). doi:10.1002/app.48396.

Alam, M. N., and Lew P. Christopher. 2018. "Natural Cellulose-Chitosan Cross-Linked Super-adsorbent Hydrogels with Superior Swelling Properties." *ACS Sustainable Chemistry and Engineering* 6 (7). doi:10.1021/acssuschemeng.8b01062.

Anitha, A., S. Sowmya, P. T. Sudheesh Kumar, S. Deepthi, K. P. Chennazhi, H. Ehrlich, M. Tsurkan, and R. Jayakumar. 2014. "Chitin and Chitosan in Selected Biomedical Applications." *Progress in Polymer Science*. doi:10.1016/j.progpolymsci.2014.02.008.

Ashkani, M., K. Kabiri, A. Salimi, H. Bouhendi, and H. Omidian. 2018. "Hybrid Hydrogel Based on Pre-Gelatinized Starch Modified with Glycidyl-Cross-linked Microgel." *Iranian Polymer Journal (English Edition)* 27 (3). doi:10.1007/s13726-018-0599-4.

Barajas-Ledesma, R. M., Antonio F. Patti, Vanessa N. L. Wong, Vikram Singh Raghuwanshi, and Gil Garnier. 2020. "Engineering Nanocellulose Super-adsorbent Structure by Controlling the Drying Rate." *Colloids and Surfaces A: Physicochemical and Engineering Aspects* 600. doi:10.1016/j.colsurfa.2020.124943.

Barajas-Ledesma, R. M., Craig W. Stocker, Vanessa N. L. Wong, Karen Little, Antonio F. Patti, and Gil Garnier. 2022. "Biodegradation of a Nanocellulose Super-adsorbent and Its Effect on the Growth of Spinach (Spinacea Oleracea)." *ACS Agricultural Science and Technology* 2 (1). doi:10.1021/acsagscitech.1c00178.

Baseer, A., Aljoscha Koenneke, Josef Zapp, Saeed A. Khan, and Marc Schneider. 2019. "Design and Characterization of Surface-Cross-linked Gelatin Nanoparticles for the Delivery of Hydrophilic Macromolecular Drugs." *Macromolecular Chemistry and Physics* 220 (18). doi:10.1002/macp.201900260.

Bouchtaoui, F. Z. E., El Houssaine Ablouh, Manal Mhada, Ihsane Kassem, Mohamed Hamid Salim, Salma Mouhib, Zineb Kassab, Houssine Sehaqui, and Mounir El Achaby. 2022. "Methylcellulose/Lignin Biocomposite as an Eco-Friendly and Multifunctional Coating Material for Slow-Release Fertilizers: Effect on Nutrients Management and Wheat Growth." *International Journal of Biological Macromolecules* 221. doi:10.1016/j.ijbiomac.2022.08.194.

Buchmann, C., Zacharias Steinmetz, Mathilde Brax, Stephan Peth, and Gabriele Ellen Schaumann. 2020. "Effect of Matric Potential and Soil-Water-Hydrogel Interactions on Biohydrogel-Induced Soil Microstructural Stability." *Geoderma* 362. doi:10.1016/j.geoderma.2019.114142.

Capanema, N. S. V., Alexandra A. P. Mansur, Anderson C. de Jesus, Sandhra M. Carvalho, Luiz C. de Oliveira, and Herman S. Mansur. 2018. "Super-adsorbent Cross-linked Carboxymethyl Cellulose-PEG Hydrogels for Potential Wound Dressing Applications." *International Journal of Biological Macromolecules* 106. doi:10.1016/j.ijbiomac.2017.08.124.

Chang, L., Liju Xu, Yaohu Liu, and Dong Qiu. 2021. "Super-adsorbent Polymers Used for Agricultural Water Retention." *Polymer Testing* 94. doi:10.1016/j.polymertesting.2020.107021.

Chaudhary, J., Sourbh Thakur, Minaxi Sharma, Vijai Kumar Gupta, and Vijay Kumar Thakur. 2020. "Development of Biodegradable Agar-Agar/Gelatin-Based Super-adsorbent Hydrogel as an Efficient Moisture-Retaining Agent." *Biomolecules* 10 (6). doi:10.3390/biom10060939.

Chaudhuri, S. D., Ayan Dey, Shweta Upganlawar, and Debabrata Chakrabarty. 2022. "Influence of Clay Concentration on the Absorption and Rheological Attributes of Modified Cellulose/Acrylic Acid Based Hydrogel and the Application of Such

Hydrogel." *Materials Chemistry and Physics* 282. doi:10.1016/j.matchemphys.2022.125942.

Chen, L., Rui Dai, Zhihua Shan, and Hui Chen. 2020. "Fabrication and Characterization of One High-Hygroscopicity Liquid Starch-Based Mulching Materials for Facilitating the Growth of Plant." *Carbohydrate Polymers* 230. doi:10.1016/j.carbpol.2019.115582.

Cheng, S., Wei Zeng, Xiaomei Liu, Junji Zhao, Xueyan Qiu, and Ziqiang Lei. 2020. "Anti-Evaporation Performance of Water in Soil of Super-adsorbent Resin with Fast Water Absorption Rate." *Water, Air, and Soil Pollution* 231 (6). doi:10.1007/s11270-020-04679-8.

Cheng, S., Xiaomei Liu, Jianghong Zhen, and Ziqiang Lei. 2019. "Preparation of Super-adsorbent Resin with Fast Water Absorption Rate Based on Hydroxymethyl Cellulose Sodium and Its Application." *Carbohydrate Polymers* 225. doi:10.1016/j.carbpol.2019.115214.

Demitri, C., F. Scalera, M. Madaghiele, A. Sannino, and A. Maffezzoli. 2013. "Potential of Cellulose-Based Super-adsorbent Hydrogels as Water Reservoir in Agriculture." *International Journal of Polymer Science* 2013. doi:10.1155/2013/435073.

Dwivedi, C., Abhishek Gupta, Abhishek Chaudhary, and Chayan Kanti Nandi. 2014. "Gold Nanoparticle Chitosan Composite Hydrogel Beads Show Efficient Removal of Methyl Parathion from Waste Water." *RSC Advances* 4 (75). doi:10.1039/c4ra03870c.

Elhassani, C. E., Younes Essamlali, Meryem Aqlil, Annie Moussemba Nzenguet, Ikram Ganetri, and Mohamed Zahouily. 2019. "Urea-Impregnated HAP Encapsulated by Lignocellulosic Biomass-Extruded Composites: A Novel Slow-Release Fertilizer." *Environmental Technology and Innovation* 15. doi:10.1016/j.eti.2019.100403.

Fang, S., Guangjian Wang, Pengcheng Li, Ronge Xing, Song Liu, Yukun Qin, Huahua Yu, Xiaolin Chen, and Kecheng Li. 2018. "Synthesis of Chitosan Derivative Graft Acrylic Acid Super-adsorbent Polymers and Its Application as Water Retaining Agent." *International Journal of Biological Macromolecules* 115. doi:10.1016/j.ijbiomac.2018.04.072.

Fang, S., Guangjian Wang, Ronge Xing, Xiaolin Chen, Song Liu, Yukun Qin, Kecheng Li, Xueqin Wang, Rongfeng Li, and Pengcheng Li. 2019. "Synthesis of Super-adsorbent Polymers Based on Chitosan Derivative Graft Acrylic Acid-Co-Acrylamide and Its Property Testing." *International Journal of Biological Macromolecules* 132. doi:10.1016/j.ijbiomac.2019.03.176.

Fekete, T., Judit Borsa, Erzsébet Takács, and László Wojnárovits. 2017. "Synthesis of Carboxymethylcellulose/Starch Super-adsorbent Hydrogels by Gamma-Irradiation." *Chemistry Central Journal* 11 (1). doi:10.1186/s13065-017-0273-5.

Gao, D., Xinfang Wei, Yukai Zhang, Yuan Ma, Guanqi Wang, Xin Zhao, Kefan Liu, Yuanfei Huo, Haiwang Wang, and Bingzhu Wang. 2021. "Preparation of TiO2-P(AM-AA) Organic-Inorganic Composite Water-Retaining Agent Based on Photocatalytic Surface-Initiated Polymerization." *Journal of Dispersion Science and Technology* 42 (4). doi:10.1080/01932691.2019.1702555.

Gharrak, A. E., Younes Essamlali, Othmane Amadine, Soumia Aboulhrouz, Anass Hafnaoui, Hakim Ghalfi, and Mohamed Zahouily. 2022. "Tunable Physicochemical Properties of Lignin and Rapeseed Oil-Based Polyurethane Coatings with Tailored

Release Property of Coated NPK Fertilizer." *Progress in Organic Coatings*. doi:10.1016/j.porgcoat.2022.106982.

Ghasri, M., Arash Jahandideh, Kourosh Kabiri, Hossein Bouhendi, Mohammad J. Zohuriaan-Mehr, and Nasrin Moini. 2019. "Glycerol-Lactic Acid Star-Shaped Oligomers as Efficient Biobased Surface Modifiers for Improving Super-adsorbent Polymer Hydrogels." *Polymers for Advanced Technologies* 30 (2). doi:10.1002/pat.4476.

Ghobashy, M. M., Basem Kh El-Damhougy, Norhan Nady, H. Abd El-Wahab, A. M. Naser, and Farag Abdelhai. 2018. "Radiation Cross-linking of Modifying Super Absorbent (Polyacrylamide/Gelatin) Hydrogel as Fertilizers Carrier and Soil Conditioner." *Journal of Polymers and the Environment* 26 (9). doi:10.1007/s10924-018-1273-9.

Guan, H. L., D. L. Yong, M. X. Fan, X. L. Yu, Z. Wang, J. J. Liu, and J. B. Li. 2018. "Sodium Humate Modified Super-adsorbent Resin with Higher Salt-Tolerating and Moisture-Resisting Capacities." *Journal of Applied Polymer Science* 135 (48). doi:10.1002/app.46892.

Irani, M., Hanafi Ismail, and Zulkifli Ahmad. 2013. "Preparation and Properties of Linear Low-Density Polyethylene-g-Poly (Acrylic Acid)/Organo-Montmorillonite Super-adsorbent Hydrogel Composites." *Polymer Testing* 32 (3). doi:10.1016/j.polymertesting.2013.01.001.

Jiang, Y., Jing Zhang, Delgado Baquerizo Manuel, Michiel Op de Beeck, Muhammad Shahbaz, Yi Chen, Xiaopeng Deng, Zhaoli Xu, Jian Li, and Zhanfeng Liu. 2022. "Rotation Cropping and Organic Fertilizer Jointly Promote Soil Health and Crop Production." *Journal of Environmental Management* 315. doi:10.1016/j.jenvman.2022.115190.

Kandasamy, G., Ravichandran Manisekaran, and Manoj Kumar Arthikala. 2024. "Chitosan Nanoplatforms in Agriculture for Multi-Potential Applications - Adsorption/Removal, Sustained Release, Sensing of Pollutants & Delivering Their Alternatives – A Comprehensive Review." *Environmental Research*. doi:10.1016/j.envres.2023.117447.

Kandil, S., El Refaie Kenawy, Azza El-Maghraby, and Elsayed H. Elashry. 2004. "Controlled Release of 2-Methyl-4-Chlorophenoxy Acetic Acid Herbicide from Waste Gelatin-Based Blends and Composites." *Journal of Applied Polymer Science* 94 (4). doi:10.1002/app.21045.

Kang, M. W., Mesenbet Yibeltal, Young Hyun Kim, Se Jin Oh, Jong Cheol Lee, Eilhann E. Kwon, and Sang Soo Lee. 2022. "Enhancement of Soil Physical Properties and Soil Water Retention with Biochar-Based Soil Amendments." *Science of the Total Environment* 836. doi:10.1016/j.scitotenv.2022.155746.

Kaur, R., Dinesh Goyal, and Shekhar Agnihotri. 2021. "Chitosan/PVA Silver Nanocomposite for Butachlor Removal: Fabrication, Characterization, Adsorption Mechanism and Isotherms." *Carbohydrate Polymers* 262. doi:10.1016/j.carbpol.2021.117906.

Kenawy, E. R., Hasnaa Hasab Elnaby, and Mohamed M. Azaam. 2024. "Synthesis of Super-adsorbent Composite Based on Chitosan-g-Poly(Acrylamide)/Attapulgite." *Polymer Bulletin* 81 (4). doi:10.1007/s00289-023-04877-4.

Kenawy, E. R., Maurizia Seggiani, Ali Hosny, Mohamed Rashad, Patrizia Cinelli, Khalil M. Saad-Allah, Mohamed El-Sharnouby, S. Shendy, and Mohamed M. Azaam. 2021. "Super-adsorbent Composites Based on Rice Husk for Agricultural Applications: Swelling Behavior, Biodegradability in Soil and Drought Alleviation." *Journal of Saudi Chemical Society* 25 (6). doi:10.1016/j.jscs.2021.101254.

Kenawy, E. R., Maurizia Seggiani, Patrizia Cinelli, Hasnaa Mohamed Hasab Elnaby, and Mohamed M. Azaam. 2020. "Swelling Capacity of Sugarcane Bagasse-g-Poly(Acrylamide)/Attapulgite Super-adsorbent Composites and Their Application as Slow Release Fertilizer." *European Polymer Journal* 133. doi:10.1016/j.eurpolymj.2020.109769.

Kenawy, E. R., Mohamed M. Azaam, and Esraa M. El-nshar. 2018. "Preparation of Carboxymethyl Cellulose-g-Poly (Acrylamide)/Montmorillonite Super-adsorbent Composite as a Slow-Release Urea Fertilizer." *Polymers for Advanced Technologies* 29 (7). doi:10.1002/pat.4315.

Kenawy, E. R., Mohamed M. Azaam, and Esraa M. El-nshar. 2019. "Sodium Alginate-g-Poly(Acrylic Acid-Co-2-Hydroxyethyl Methacrylate)/Montmorillonite Super-adsorbent Composite: Preparation, Swelling Investigation and Its Application as a Slow-Release Fertilizer." *Arabian Journal of Chemistry* 12 (6). doi:10.1016/j.arabjc.2017.10.013.

Kono, H., Sayaka Fujita, and Ikuo Oeda. 2013. "Comparative Study of Homogeneous Solvents for the Esterification Cross-linking of Cellulose with 1,2,3,4-butanetetracarboxylic Dianhydride and Water Absorbency of the Reaction Products." *Journal of Applied Polymer Science* 127 (1): 478–86. doi:10.1002/app.37736.

Lan, G., Ming Zhang, Yongqiang Liu, Haiyan Qiu, Songsong Xue, Tailiang Zhang, and Qianxia Xu. 2019. "Synthesis and Swelling Behavior of Super-Absorbent Soluble Starch-g-Poly(AM-Co-NaAMC14S) Through Graft Copolymerization and Hydrolysis." *Starch - Stärke*, February, 1800272. doi:10.1002/star.201800272.

Lawrencia, D., See Kiat Wong, Darren Yi Sern Low, Bey Hing Goh, Joo Kheng Goh, Uracha Rungsardthong Ruktanonchai, Apinan Soottitantawat, Learn Han Lee, and Siah Ying Tang. 2021. "Controlled Release Fertilizers: A Review on Coating Materials and Mechanism of Release." *Plants*. doi:10.3390/plants10020238.

Lejcuś, K., Michał Śpitalniak, and Jolanta Dabrowska. 2018. "Swelling Behaviour of Super-adsorbent Polymers for Soil Amendment under Different Loads." *Polymers* 10 (3). doi:10.3390/polym10030271.

Li, L., Mingxu Chen, Xiangming Zhou, Lingchao Lu, Yanlin Li, Chenchen Gong, and Xin Cheng. 2017. "A Case of Water Absorption and Water/Fertilizer Retention Performance of Super Absorbent Polymer Modified Sulphoaluminate Cementitious Materials." *Construction and Building Materials* 150. doi:10.1016/j.conbuildmat.2017.05.219.

Li, S. S., Song Bo Wang, Yu Chen, Qi Shun Zhu, Ling Min Lan, Huaitian Bu, Tian Hu, and Gang Biao Jiang. 2023. "Biodegradable, Anti-Freezing and Self-Healable Hydrogel Mulch Film for Weed Control." *Chemical Engineering Journal* 462. doi:10.1016/j.cej.2023.142211.

Liu, J., Shunyi Wang, Ke Xu, Zaibi Fan, Panpan Wang, Zijun Xu, Xueqin Ren, Shuwen Hu, and Zideng Gao. 2020. "Fabrication of Double Cross-linked Chitosan/Gelatin

Membranes with Na+ and PH Dual-Responsive Controlled Permeability." *Carbohydrate Polymers* 236. doi:10.1016/j.carbpol.2020.115963.

Liu, S., Qinglin Wu, Xiuxuan Sun, Yiying Yue, Brenda Tubana, Rongjie Yang, and Huai N. Cheng. 2021. "Novel Alginate-Cellulose Nanofiber-Poly(Vinyl Alcohol) Hydrogels for Carrying and Delivering Nitrogen, Phosphorus and Potassium Chemicals." *International Journal of Biological Macromolecules* 172. doi:10.1016/j.ijbiomac.2021.01.063.

Lv, Q., Min Wu, and Yong Shen. 2019. "Enhanced Swelling Ratio and Water Retention Capacity for Novel Super-Absorbent Hydrogel." *Colloids and Surfaces A: Physicochemical and Engineering Aspects* 583. doi:10.1016/j.colsurfa.2019.123972.

Meng, Y., and Lin Ye. 2017. "Synthesis and Swelling Property of Super-adsorbent Starch Grafted with Acrylic Acid/2-Acrylamido-2-Methyl-1-Propanesulfonic Acid." *Journal of the Science of Food and Agriculture* 97 (11). doi:10.1002/jsfa.8247.

Merwe, R. du Toit van der, Neill Jurgens Goosen, and Robert William Mc Clelland Pott. 2022. "Macroalgal-Derived Alginate Soil Amendments for Water Retention, Nutrient Release Rate Reduction, and Soil PH Control." *Gels* 8 (9). doi:10.3390/gels8090548.

Mikula, K., Grzegorz Izydorczyk, Dawid Skrzypczak, Małgorzata Mironiuk, Konstantinos Moustakas, Anna Witek-Krowiak, and Katarzyna Chojnacka. 2020. "Controlled Release Micronutrient Fertilizers for Precision Agriculture – A Review." *Science of the Total Environment* 712. doi:10.1016/j.scitotenv.2019.136365.

Motamedi, E., Babak Motesharezedeh, Ahmad Shirinfekr, and Seyed Mahmood Samar. 2020. "Synthesis and Swelling Behavior of Environmentally Friendly Starch-Based Super-adsorbent Hydrogels Reinforced with Natural Char Nano/Micro Particles." *Journal of Environmental Chemical Engineering* 8 (1). doi:10.1016/j.jece.2019.103583.

Motamedi, Elaheh, Marzieh Safari, and Mehri Salimi. 2023. "Improvement of Tomato Yield and Quality Using Slow Release NPK Fertilizers Prepared by Carnauba Wax Emulsion, Starch-Based Latex and Hydrogel Nanocomposite Combination." *Scientific Reports* 13 (1). doi:10.1038/s41598-023-38445-7.

Mushtaq, F., Zulfiqar Ali Raza, Syeda Rubab Batool, Muhammad Zahid, Ozgun Can Onder, Ammara Rafique, and Muhammad Anwaar Nazeer. 2022. "Preparation, Properties, and Applications of Gelatin-Based Hydrogels (GHs) in the Environmental, Technological, and Biomedical Sectors." *International Journal of Biological Macromolecules*. doi:10.1016/j.ijbiomac.2022.07.168.

Narayanan, A., Ravishankar Kartik, Elanchezhian Sangeetha, and Raghavachari Dhamodharan. 2018. "Super Water Absorbing Polymeric Gel from Chitosan, Citric Acid and Urea: Synthesis and Mechanism of Water Absorption." *Carbohydrate Polymers* 191. doi:10.1016/j.carbpol.2018.03.028.

Olad, A., Fatemeh Doustdar, and Hamed Gharekhani. 2018. "Starch-Based Semi-IPN Hydrogel Nanocomposite Integrated with Clinoptilolite: Preparation and Swelling Kinetic Study." *Carbohydrate Polymers* 200. doi:10.1016/j.carbpol.2018.08.014.

Olad, A., Mahyar Pourkhiyabi, Hamed Gharekhani, and Fatemeh Doustdar. 2018. "Semi-IPN Super-adsorbent Nanocomposite Based on Sodium Alginate and Montmorillonite: Reaction Parameters and Swelling Characteristics." *Carbohydrate Polymers* 190. doi:10.1016/j.carbpol.2018.02.088.

Pourjavadi, A., Mohammad Sadeghi, Mohammad Mahmodi Hashemi, and Hossein Hosseinzadeh. 2006. "Synthesis and Absorbency of Gelatin-Graft-Poly(Sodium Acrylate-Co- Acrylamide) Super-adsorbent Hydrogel with Saltand PH-Responsiveness Properties." *E-Polymers*. doi:10.1515/epoly.2006.6.1.728.

Pulat, M., and Nilgün Uğurlu. 2016. "Preparation and Characterization of Biodegradable Gelatin-PAAm-Based IPN Hydrogels for Controlled Release of Maleic Acid to Improve the Solubility of Phosphate Fertilizers." *Soft Materials* 14 (4). doi:10.1080/1539445X.2016.1181088.

Qi, Taomei, S. L., Shao Fei Zhang, Xiao Bai, Jiao Chen, Mengjie Huang, and Mingzhu Liu. 2020. "Zein Coated Porous Carboxymethyl Starch Fertilizer for Iron Promoting and Phosphate Sustainable Release." *Journal of Cleaner Production* 258. doi:10.1016/j.jclepro.2020.120778.

Qiao, D., Hongsheng Liu, Long Yu, Xianyang Bao, George P. Simon, Eustathios Petinakis, and Ling Chen. 2016. "Preparation and Characterization of Slow-Release Fertilizer Encapsulated by Starch-Based Super-adsorbent Polymer." *Carbohydrate Polymers* 147. doi:10.1016/j.carbpol.2016.04.010.

Qureshi, M. A., Nahid Nishat, Sapana Jadoun, and Mohd Zaid Ansari. 2020. "Polysaccharide Based Super-adsorbent Hydrogels and Their Methods of Synthesis: A Review." *Carbohydrate Polymer Technologies and Applications*. doi:10.1016/j.carpta.2020.100014.

Rahmanifar, B., and Shahram Moradi Dehaghi. 2014. "Removal of Organochlorine Pesticides by Chitosan Loaded with Silver Oxide Nanoparticles from Water." *Clean Technologies and Environmental Policy* 16 (8). doi:10.1007/s10098-013-0692-5.

Rather, R. A., Mushtaq A. Bhat, and Aabid H. Shalla. 2022. "An Insight into Synthetic and Physiological Aspects of Super-adsorbent Hydrogels Based on Carbohydrate Type Polymers for Various Applications: A Review." *Carbohydrate Polymer Technologies and Applications*. doi:10.1016/j.carpta.2022.100202.

Rattanamanee, A., Hataichanoke Niamsup, La ongnuan Srisombat, Ruangsri Watanesk, and Surasak Watanesk. 2013. "Effect of Gelatin on Secondary Structure, Crystallinity and Swelling Behavior of Silk Fibroin-Gelatin Hydrogelsand Its Application in Controlledrelease of Nitrogen." In *Advanced Materials Research*. Vol. 750–752. doi:10.4028/www.scientific.net/AMR.750-752.1347.

Rattanamanee, A., Hataichanoke Niamsup, La ongnuan Srisombat, Winita Punyodom, Ruangsri Watanesk, and Surasak Watanesk. 2015. "Role of Chitosan on Some Physical Properties and the Urea Controlled Release of the Silk Fibroin/Gelatin Hydrogel." *Journal of Polymers and the Environment* 23 (3). doi:10.1007/s10924-014-0703-6.

Roy, S., and Jong Whan Rhim. 2020. "Preparation of Antimicrobial and Antioxidant Gelatin/Curcumin Composite Films for Active Food Packaging Application." *Colloids and Surfaces B: Biointerfaces* 188. doi:10.1016/j.colsurfb.2019.110761.

Saelim, T., Sermpong Sairiam, Krisana Siralertmukul, Supawi Watcharamul, and Roongkan Nuisin. 2020. "Chitosan Beads as the Adsorbent for Glyphosate Removal from an Aqueous Solution." *Journal of Metals, Materials and Minerals* 30 (3). doi:10.14456/jmmm.2020.40.

Saifuddin, N., C. Y. Nian, L. W. Zhan, and K. X. Ning. 2011. "Chitosan-Silver Nanoparticles Composite as Point-of-Use Drinking Water Filtration System for Household to Remove Pesticides in Water." *Asian Journal of Biochemistry* 6 (2). doi:10.3923/ajb.2011.142.159.

Salimi, M., Badr eddine Channab, Ayoub El Idrissi, Mohamed Zahouily, and Elaheh Motamedi. 2023. "A Comprehensive Review on Starch: Structure, Modification, and Applications in Slow/Controlled-Release Fertilizers in Agriculture." *Carbohydrate Polymers*. doi:10.1016/j.carbpol.2023.121326.

Salimi, M., Elaheh Motamedi, Marzieh Safari, and Babak Motesharezadeh. 2021. "Synthesis of Urea Slow-Release Fertilizer Using a Novel Starch-g-Poly(Styrene-Co-Butylacrylate) Nanocomposite Latex and Its Impact on a Model Crop Production in Greenhouse." *Journal of Cleaner Production* 322. doi:10.1016/j.jclepro.2021.129082.

Sarkar, A., Dipak Ranjan Biswas, Samar Chandra Datta, Brahma Swarup Dwivedi, Ranjan Bhattacharyya, Rajesh Kumar, Kali Kinkar Bandyopadhyay, et al., 2021. "Preparation of Novel Biodegradable Starch/Poly(Vinyl Alcohol)/Bentonite Grafted Polymeric Films for Fertilizer Encapsulation." *Carbohydrate Polymers* 259. doi:10.1016/j.carbpol.2021.117679.

Sarmah, D., and Niranjan Karak. 2020. "Biodegradable Super-adsorbent Hydrogel for Water Holding in Soil and Controlled-Release Fertilizer." *Journal of Applied Polymer Science* 137 (13). doi:10.1002/app.48495.

Sharma, R., Jaya Bajpai, A. K. Bajpai, Somen Acharya, R. B. Shrivastava, and S. K. Shukla. 2014. "Designing Slow Water-Releasing Alginate Nanoreserviors for Sustained Irrigation in Scanty Rainfall Areas." *Carbohydrate Polymers* 102 (1). doi:10.1016/j.carbpol.2013.11.059.

Situ, Y., Cailan Huang, Yifan Yang, Zongwen Liao, Xiaoyun Mao, and Xian chen. 2023. "Synthesis and Application of Super Absorbent Polymers Synthesized with Ammonia Solution and Diatomaceous Earth with Low Toxic Residues." *Environmental Technology and Innovation* 32. doi:10.1016/j.eti.2023.103371.

Song, J., Haidong Zhao, Guizhe Zhao, Yang Xiang, and Yaqing Liu. 2019. "Novel Semi-IPN Nanocomposites with Functions of Both Nutrient Slow-Release and Water Retention. 1. Microscopic Structure, Water Absorbency, and Degradation Performance." *Journal of Agricultural and Food Chemistry* 67 (27). doi:10.1021/acs.jafc.9b00888.

Songara, J. C., Jayantilal N. Patel, and Alka A. Mungray. 2022. "Preparation and Characterization of PAA/GG-Zeolite Nano-Composite Hydrogel for Agricultural Applications." *Journal of the Indian Chemical Society* 99 (10). doi:10.1016/j.jics.2022.100686.

Su, X., and Biqiong Chen. 2018. "Tough, Resilient and PH-Sensitive Interpenetrating Polyacrylamide/Alginate/Montmorillonite Nanocomposite Hydrogels." *Carbohydrate Polymers* 197. doi:10.1016/j.carbpol.2018.05.082.

Swain, S. K., Bhargabi Shur, and Subrata K. Patra. 2013. "Poly(Acrylamide-Co-Vinyl Alcohol)-Super-adsorbent Materials Reinforced by Modified Clay." *Polymer Composites* 34 (11). doi:10.1002/pc.22583.

Tiamwong, S., Pratchayaporn Yukhajon, Pittayagorn Noisong, Maliwan Subsadsana, and Sira Sansuk. 2023. "Eco-Friendly Starch Composite Supramolecular Alginate–Ca2+

Hydrogel as Controlled-Release P Fertilizer with Low Responsiveness to Multiple Environmental Stimuli." *Gels* 9 (3). doi:10.3390/gels9030204.

Tian, H., Zeli Li, Panfang Lu, Yong Wang, Cong Jia, Huaili Wang, Zhiguang Liu, and Min Zhang. 2021. "Starch and Castor Oil Mutually Modified, Cross-Linked Polyurethane for Improving the Controlled Release of Urea." *Carbohydrate Polymers* 251. doi:10.1016/j.carbpol.2020.117060.

Vieira, T., Valter Antonio Becegato, and Alexandre Tadeu Paulino. 2021. "Equilibrium Isotherms, Kinetics, and Thermodynamics of the Adsorption of 2,4-Dichlorophenoxyacetic Acid to Chitosan-Based Hydrogels." *Water, Air, & Soil Pollution* 232 (2): 60. doi:10.1007/s11270-021-05021-6.

Vu, C. T., and Tingting Wu. 2022. "MgAl-Layered Double Hydroxides/Sodium Alginate Beads for Nitrate Adsorption from Groundwater and Potential Use as a Slow-Release Fertilizer." *Journal of Cleaner Production* 379. doi:10.1016/j.jclepro.2022.134508.

Wang, N., Bing Wang, Yongshan Wan, Bin Gao, and Vishnu D. Rajput. 2023. "Alginate-Based Composites as Novel Soil Conditioners for Sustainable Applications in Agriculture: A Critical Review." *Journal of Environmental Management*. doi:10.1016/j.jenvman.2023.119133.

Wang, W., Zhengli Yang, Aiping Zhang, and Shiqi Yang. 2021. "Water Retention and Fertilizer Slow Release Integrated Super-adsorbent Synthesized from Millet Straw and Applied in Agriculture." *Industrial Crops and Products* 160. doi:10.1016/j.indcrop.2020.113126.

Wang, Z., Aimin Ning, Puhui Xie, Guangqin Gao, Lixia Xie, Xin Li, and Andong Song. 2017. "Synthesis and Swelling Behaviors of Carboxymethyl Cellulose-Based Super-adsorbent Resin Hybridized with Graphene Oxide." *Carbohydrate Polymers* 157. doi:10.1016/j.carbpol.2016.09.070.

Wei, J., Haiyuan Yang, Hui Cao, and Tianwei Tan. 2016. "Using Polyaspartic Acid Hydro-Gel as Water Retaining Agent and Its Effect on Plants under Drought Stress." *Saudi Journal of Biological Sciences* 23 (5). doi:10.1016/j.sjbs.2015.08.016.

Wen, P., Zhansheng Wu, Yanhui He, Bang Ce Ye, Yajie Han, Jun Wang, and Xinyuan Guan. 2016. "Microwave-Assisted Synthesis of a Semi-Interpenetrating Polymer Network Slow-Release Nitrogen Fertilizer with Water Absorbency from Cotton Stalks." *ACS Sustainable Chemistry and Engineering* 4 (12). doi:10.1021/acssuschemeng.6b01466.

Xiang, Y., Xudong Ru, Jinguo Shi, Jiang Song, Haidong Zhao, Yaqing Liu, Dongdong Guo, and Xin Lu. 2017. "Preparation and Properties of a Novel Semi-IPN Slow-Release Fertilizer with the Function of Water Retention." *Journal of Agricultural and Food Chemistry* 65 (50). doi:10.1021/acs.jafc.7b03827.

Xiao, X., Long Yu, Fengwei Xie, Xianyang Bao, Hongsheng Liu, Zhili Ji, and Ling Chen. 2017. "One-Step Method to Prepare Starch-Based Super-adsorbent Polymer for Slow Release of Fertilizer." *Chemical Engineering Journal* 309. doi:10.1016/j cej.2016.10.101.

Xiong, H., Hui Peng, Xi'e Ye, Yanrong Kong, Na Wang, Fenghong Yang, Ben Hur Meni, and Ziqiang Lei. 2022. "High Salt Tolerance Hydrogel Prepared of Hydroxyethyl Starch and Its Ability to Increase Soil Water Holding Capacity and Decrease Water Evaporation." *Soil and Tillage Research* 222. doi:10.1016/j.still.2022.105427.

Yang, D. 2022. "Recent Advances in Hydrogels." *Chemistry of Materials*. doi:10.1021/acs.chemmater.2c00188.

Yang, G., Hongmeng Zhao, Yanli Liu, Zeli Li, Feng Gao, Qiang Zhang, Peng Zou, Zhiguang Liu, and Min Zhang. 2023. "Slow Release Fertilizers Based on Polyphosphate/Montmorillonite Nanocomposites for Improving Crop Yield." *Arabian Journal of Chemistry* 16 (7). doi:10.1016/j.arabjc.2023.104871.

Yang, H., Fengli Dai, Hongxia Chen, Yufeng He, Zeyuan Wang, and Rongmin Wang. 2023. "Fabrication of Stalk Fiber/Geopolymers-Based Slow-Release Fertilizer with Agricultural Waste and Loess for Promoting Plant Growth." *Journal of Environmental Chemical Engineering* 11 (2). doi:10.1016/j.jece.2023.109481.

Yu, X., Zhe Wang, Jiajun Liu, Hua Mei, Dingli Yong, and Junbo Li. 2019. "Preparation, Swelling Behaviors and Fertilizer-Release Properties of Sodium Humate Modified Super-adsorbent Resin." *Materials Today Communications* 19. doi:10.1016/j.mtcomm.2018.12.015.

Zhang, M., and Jisheng Yang. 2021. "Preparation and Characterization of Multifunctional Slow Release Fertilizer Coated with Cellulose Derivatives." *International Journal of Polymeric Materials and Polymeric Biomaterials* 70 (11). doi:10.1080/00914037.2020.1765352.

Zhang, S., Furui He, Xiuqin Fang, Xinyu Zhao, Yuanyuan Liu, Gaobo Yu, Yang Zhou, Yuhong Feng, and Jiacheng Li. 2022. "Enhancing Soil Aggregation and Acetamiprid Adsorption by Ecofriendly Polysaccharides Hydrogel Based on Ca2+- Amphiphilic Sodium Alginate." *Journal of Environmental Sciences (China)* 113. doi:10.1016/j.jes.2021.05.042.

Zhang, S., Wei Wang, Haiyan Wang, Wenyuan Qi, Ling Yue, and Qingfu Ye. 2014. "Synthesis and Characterisation of Starch Grafted Super-adsorbent via 10 MeV Electron-Beam Irradiation." *Carbohydrate Polymers* 101 (1). Elsevier: 798–803. doi:10.1016/J.CARBPOL.2013.10.009.

Zhang, W., Xiang Liao Sun, Qian Yang, Yuanyuan Guo, Yanjun Cui, Yongsheng Xiang, Bing Hu, Jia Wei, and Peng Tu. 2024. "*In Situ* Forming of PEG-NH2/Dialdehyde Starch Schiff-Base Hydrogels and Their Application in Slow-Release Urea." *International Journal of Biological Macromolecules* 256. doi:10.1016/j.ijbiomac.2023.128355.

Zhang, Z., Noureddine Abidi, Lucian Lucia, Sakineh Chabi, Christian T. Denny, Prakash Parajuli, and Shaida Sultana Rumi. 2023. "Cellulose/Nanocellulose Super-adsorbent Hydrogels as a Sustainable Platform for Materials Applications: A Mini-Review and Perspective." *Carbohydrate Polymers*. doi:10.1016/j.carbpol.2022.120140.

Zhao, C., Min Zhang, Zhiguang Liu, Yanle Guo, and Qiang Zhang. 2019. "Salt-Tolerant Super-adsorbent Polymer with High Capacity of Water-Nutrient Retention Derived from Sulfamic Acid-Modified Starch." *ACS Omega* 4 (3). doi:10.1021/acsomega.9b00486.

Biographical Sketches

El-Refaie Kenawy

Affiliation:

Polymer Research Group, Department of Chemistry, Faculty of Science, Tanta University.

Education: Professor of Polymer Chemistry from 2004.

Research and Professional Experience:

1. Head of Chemistry Department, Faculty of Science, Tanta University, EGYPT, (2020- 2021) and (2014-2017).
2. Vice Dean for Postgraduate Studies and Research, Faculty of Science, Tanta University, (2017- 2020).
3. Director of Technology Transfer Office at Tanta University, Egypt (2013-2018).
4. Professor at Chemistry Department, Faculty of Science, King Saud University, Kingdom of Saudi Arabia 2008-2010; Director of Polymer and Nanofibers Research Group, KSA.
5. Visiting Professor at Tuskegee University, Alabama, USA March 2006- Feb 28th 2007
6. Visiting Professor at Tokyo Institute of Technology, Tokyo, Japan July 2006-Feb. 2007.
7. Visiting Professor at the Chemical Engineering Dept., Virginia Commonwealth University (VCU) Richmond, USA (June 2000 to June 2002).
8. Post-Doctor Fellow at University of Pisa, Italy, supported by ICS-UNIDO Italy (October 1997-September 1998).
9. Post-Doctor Fellow, University of Gent, Belgium, (January 1993 till August 1994).

Professional Appointments:

1. MOU mediator and coordinator between University of Tanta, Egypt and National Research Center, Egypt 2020.

2. MOU mediator and coordinator between University of Tanta, Egypt and City of Scientific Research and Technology, Alexandria, Egypt 2007.
3. MOU between King Saud University, KSA, and Bayreuth University, Germany to initiate a Joint degree for Master in Polymer Science (In progress) 2010.
4. Head of Scientific committee for interviewing students and faculty members in science applying to PhD missions and Post doctors from Ministry of higher education, mission department since 2012 till now.
5. Member of the promotion committee of the professors and associate professors for Egyptians Universities (Supreme University Council) in Organic Chemistry (No 13) from 2019 to 2022.
6. Member of the promotion committee of the professors and associate professors for Egyptian Atomic Energy, Chemical science, 2020-2023.

Honors:

1. Abdul Hameed Shoman Award for Young Arab Scientists in Chemistry 2004.
2. Misr Elkher Award (Egypt) for the most cited scientist 2012.
3. Tanta University Award Appreciation in Basic Science 2014.
4. Excellence Award in University Performance 2016.
5. Tanta University Award for the most cited scientist 2019.
6. Tanta University Research Awards.
7. King Saud University Prize for distinguished Research articles, 2009.
8. Prize from Tanta University for 50 International publications from the year 2011-2021.

Publications from the Last 3 Years:

[1] Kenawy, E. R., Tarek B. Edries, Fatma S. Abouharga, and Youssef I. Moharram. "Electrochemical Analysis of Polyvinyl Alcohol/Magnetic Iron Oxide Composite Nanofibers for Supercapacitor Applications." *Egyptian Journal of Chemistry* 67, no. 4 (2024): 215-224.

[2] Kenawy, E. R., A. M. Abd El. Hay, Nermeen Saad, Mieohamed M. Azaam, and Kamel R. Shoueir. "Synthesis, characterization of poly l (+) lactic acid and its application in sustained release of isosorbide dinitrate." *Scientific Reports* 14, no. 1 (2024): 7062.

[3] Kenawy, E. R., Youssef I. Moharram, Fatma S. Abouharga, and Mona Elfiky. "Electrospun network based on polyacrylonitrile-polyphenyl/ titanium oxide nanofibers for high-performance supercapacitor device." *Scientific Reports* 14, no. 1 (2024): 6683.

[4] El-Deeb, M., Maha M. Ismail, El-Refaie Kenawy, and Nour A. Habib. "Evaluation of antibacterial activity of an experimental dental adhesive containing synthesized quaternary ammonium compound: *in vitro* study." *Discover Applied Sciences* 6, no. 3 (2024): 120.

[5] Ghazy, A. R., Diana F. Abotalb, El-Refaie Kenawy, R. Ghazy, and Sayed A. Abdel Gawad. "Optical, photo-physical and photo-stability characterizations of malachite green as a laser dye, combined with TD-DFT simulations." *Optical and Quantum Electronics* 56, no. 3 (2024): 377.

[6] AlEidan, K. E., El-Refaie Kenawy, Hamada SA Mandour, Mohamed M. Azaam, and Bander S. AlOtaibi. "Promoting the strength of Impact Copolymer Polypropylene (ICP) via innovative compounding approaches." *Egyptian Journal of Chemistry* (2024).

[7] El-Zawawy, N. A., El-Refaie Kenawy, Sara Ahmed, and Shimaa El-Sapagh. "Bioproduction and optimization of newly characterized melanin pigment from Streptomyces djakartensis NSS-3 with its anticancer, antimicrobial, and radioprotective properties." *Microbial Cell Factories* 23, no. 1 (2024): 23.

[8] Kenawy, E. R., Mohammed SA El-Moaty, Mamdooh Ghoneum, Hesham MA Soliman, Ahmed A. El-Shanshory, and S. Shendy. "Biobran-loaded core/shell nanofibrous scaffold: a promising wound dressing candidate." *RSC advances* 14, no. 7 (2024): 4930-4945.

[9] El Zawawy, N. A., Samy El-Safty, El-Refaie Kenawy, Sara Ibrahim Salem, Sameh S. Ali, and Yehia A-G. Mahmoud. "Exploring the biomedical potential of a novel modified glass ionomer cement against the pandrug-resistant oral pathogen Candida albicans SYN-01." *Journal of Oral Microbiology* 15, no. 1 (2023): 2195741.

[10] Mohammed, A. A. A., Patrizia Cinelli, and El-Refaie Kenawy. "Structural and spectroscopic features of naphthol green B integrated for improved light harvesting capability of organic/inorganic hybrid solar cells." *Optical Materials* 146 (2023): 114516.

[11] Kenawy, E. R., Mahmoud A. El-Meligy, Zeinab S. Ghaly, Marwa E. Kenawy, and Elbadawy A. Kamoun. "Novel Physically-Crosslinked Caffeine and Vitamin C-Loaded PVA/Aloe Vera Hydrogel Membranes for Topical Wound Healing: Synthesis, Characterization and *In-Vivo* Wound Healing Tests." *Journal of Polymers and the Environment* (2023): 1-18.

[12] Kenawy, E. R., Samar A. Khattab, Heikki Tenhu, and Mohamed M. Azaam. "Ammonium salts of microcrystalline cellulose-g-poly (acrylonitrile): toxicity, antioxidant and anti-inflammatory properties." *Cellulose* (2023): 1-16.

[13] Kenawy, E. R., Sahar H. El-Khalafy, Hamed A. Abosharaf, Esraa M. El-nshar, Ahmed R. Ghazy, and Mohamed M. Azaam. "Synthesis, Characterization, and Anticancer Potency of Branched Poly (p-Hydroxy Styrene) Schiff-Bases." *Macromolecular Bioscience* 23, no. 11 (2023): 2300090.

[14] Kenawy, E. R., Ahmed R. Ghazy, Hala F. Rizk, and S. Shendy. "Microwave-assisted of new derivatives of polyimine conjugated polymer based on Schiff base: synthesis, characterization, and photo-physical properties as a photoluminescent materials." *Scientific Reports 13*, no. 1 (2023): 18686.

[15] Ali, N., El-Refaie Kenawy, A. A. Wadoud, and M. I. Elhadary. "Wearable Electrospun Piezoelectric Mats Based on a PVDF Nanofiber–ZnO@ ZnS Core–Shell Nanoparticles Composite for Power Generation." Nanomaterials 13, no. 21 (2023): 2833. Salem, M. M., El-Refaie Kenawy, Hesham MH Zakaly, Antoaneta Ene, Mohamed M. Azaam, Tarek B. Edries, Di Zhou et al., "Electrospun PVDF/Barium hexaferrite fiber composites for enhanced electromagnetic shielding in the X-band range." *Results in Physics* 53 (2023): 106975.

[16] Kenawy, E. R., Ahmed R. Ghazy, Ahmed F. Al-Hossainy, Mohamed Bishr, and Mohamed M. Azzam. "Synthesis, electrospinning, and molecular docking of poly (methyl methacrylate) Schiff bases and their applications as polymeric antimicrobial agents and for dye removal." *Environmental Science and Pollution Research* (2023): 1-16.

[17] Yadav, N., Jagriti Narang, Anil Kumar Chhillar, Jogender Singh Rana, Mohd Usman Mohd Siddique, El-Refaie Kenawy, Saad Alkahtani, Mohd Neyaz Ahsan, Amit Kumar Nayak, and Md Saquib Hasnain. "Diagnostic methods employing kidney biomarkers clinching biosensors as promising tool." *Sensors International* (2023): 100253.

[18] Mansour, H., Samia M. El-signey, Mohamed Ouf, El-Refaie S. Kenawy, and Kamel R. Shoueir. "Bio-extract inspired green synthesis of AgNPs amalgamated polyurethane nanofiber promoting photocatalytic decomposition of atrazine and antibacterial activity." *Biomass Conversion and Biorefinery* (2023): 1-14.

[19] El-Said, K., Abd El-hamid, and El-Refaie Kenawy. "Peanut peel extract inhibits colorectal cell lines proliferation by apoptotic mechanism." *Biological and Biomedical Journal* 1, no. 1 (2023): 11-16.

[20] El-Khalafy, S. H., Mahmoud T. Hassanein, Samia M. Elsigeny, Hazem F. Taha, Kamel R. Shoueir, and El-Refaie S. Kenawy. "Catalytic activity of Mn (III) porphyrin complex supported onto cross linked polymers in the green oxidation of malathion with hydrogen peroxide in aqueous solution." *Arabian Journal of Chemistry* 16, no. 8 (2023): 104969.

[21] El-Sheekh, M. M., Wesam E. Yousuf, El-Refaie Kenawy, and Tarek M. Mohamed. "Biosynthesis of cellulose from Ulva lactuca, manufacture of nanocellulose and its application as antimicrobial polymer." *Scientific Reports* 13, no. 1 (2023): 10188.

[22] Elmaghraby, N. A., Ahmed M. Omer, El-Refaie Kenawy, Mohamed Gaber, Mohamed A. Hassaan, Safaa Ragab, Ismail Hossain, and Ahmed El Nemr. "Electrospun cellulose acetate/activated carbon composite modified by EDTA (rC/AC-EDTA) for efficient methylene blue dye removal." *Scientific Reports* 13, no. 1 (2023): 9919.

[23] Kenawy, E. R., Hasnaa Hasab Elnaby, and Mohamed M. Azaam. "Synthesis of superabsorbent composite based on chitosan-g-poly (acrylamide)/attapulgite." Polymer Bulletin (2023): 1-17.

[24] Kenawy, E. R., Samar Abdelhady, and Mohamed M. Azaam. "Chemical modification, electrospinning and biological activities of pluronic F68." *Polymer Bulletin* 80, no. 5 (2023): 5725-5740.

[25] Elshenawy, E. A., Manal Ahmed El-Ebiary, El-Refaie Kenawy, and Gehan Abdelmonem El-Olimy. "Modification of glass-ionomer cement properties by quaternized chitosan-coated nanoparticles." *Odontology* 111, no. 2 (2023): 328-341.

[26] Kenawy, E. R., Elbadawy A. Kamoun, Samia M. Elsigeny, Samira Haikal, Ashraf A. El-Shehawy, and Yehia AG Mahmoud. "Physically cross-linked PVA-quaternized chitosan-Ag NPs composite hydrogel membranes for potential topical wound healing applications: Synthesis, physicochemical properties, and *in vitro* bioevaluation." *Journal of Applied Pharmaceutical Science* 13, no. 3 (2023): 023-035.

[27] Elmaghraby, N. A., Ahmed M. Omer, El-Refaie Kenawy, Mohamed Gaber, Safaa Ragab, and Ahmed El Nemr. "Composite nanofiber formation using a mixture of cellulose acetate and activated carbon for oil spill treatment." *Environmental Science and Pollution Research* 30, no. 13 (2023): 38683-38699.

[28] El-Khalafy, S. H., Mohamed M. Azaam, Esraa M. El-nshar, Elbadawy A. Kamoun, and El-Refaie Kenawy. "Catalytic activity of Co (II)-porphyrin anchored onto polymeric support of electrospun polyacrylonitrile nanofiber: synthesis and efficient green oxidation of crystal violet dye with hydrogen peroxide." *Biomass Conversion and Biorefinery* (2023): 1-12.

[29] Kenawy, E. R., Elbadawy A. Kamoun, Mahmoud Serag Eldin, Hesham MA Soliman, Shahira H. EL-Moslamy, Esmail M. El-Fakharany, and Abdel-baset M. Shokr. "Electrospun PVA–dextran nanofibrous scaffolds for acceleration of topical wound healing: nanofiber optimization, characterization and *in vitro* assessment." *Arabian Journal for Science and Engineering* 48, no. 1 (2023): 205-222.

[30] Kenawy, E. R., Mohamed Rashad, Ali Hosny, Samah Shendy, Dina Gad and Khalil M. Saad-Allah, Enhancement of growth and physiological traits under drought stress in Faba bean (*Vicia faba* L.) using nanocomposite, *Journal of Plant Interactions* (2022) 17 (1), 404-418.

[31] Elmaghraby, N. A., Ahmed M Omer, El-Refaie Kenawy, Mohamed Gaber, Ahmed El Nemr, Fabrication of cellulose acetate/cellulose nitrate/carbon black nanofiber composite for oil spill treatment, *journal of Biomass Conversion and Biorefinery,* (2022), 1-19.

[32] Kenawy, E. R., Mohamed Abd AlRaheem Abu Saied, Ahmed Saafan, and Sarah Salem. "Synthesis, characterization and Applications of Novel modified chitosan in Water Treatment." *Egyptian Journal of Chemistry* 65, no. 10 (2022): 561-575.

[33] El-Nagar, I., Ahmed M Youssef, AA Abd El-Hakim, El-Refaie Kenawy, Hamada SA Mandour, Tawfik A Khattab, Novel Hydrazone Chromophore Sensor for Metallochromic Determination of Cadmium Ions, *Chemosensors journal* (2022) 10 (11), 451.

[34] El-Gharably, A. A., El-Refaie S Kenawy, Ahmed A Safaan, Saad A Aboamna, Yehia AG Mahmoud, Synthesis, characterization, antimicrobial and anticancer evaluation of N-aryl aminochitosan, *Polymer Bulletin* (2022) 79 (11), 9925-9939.

[35] Kenawy, E. R., Mohamed Abd AlRaheem Abu Saied, Ahmed Saafan, Sarah Salem, Synthesis, characterization and Applications of Novel modified chitosan in Water Treatment. *Egyptian Journal of Chemistry* (2022) 65 (10), 561-575.

[36] Elshenawy, E. A., Manal Ahmed El-Ebiary, El-Refaie Kenawy, Gehan Abdelmonem El-Olimy, Modification of glass-ionomer cement properties by quaternized chitosan-coated nanoparticles, *Odontology,* (2022) 1-14.

[37] Alasmary, F. A., El-Refaie Kenawy, Nehal M El-Deeb, Elbadawy A Kamoun, Samar A Khattab, Abdulnasser Mahmoud Karami, Patrizia Cinelli, Mohamed M Azaam, Synthesis, antimicrobial and anticancer activities of Tetronic 1107 Schiff bases, *Polymers for Advanced Technologies* (2022) 33 (9), 2787-2797.

[38] Kenawy, E. R., Ali Ibrahim & Ahmed F. Al-Hossainy, Study of the structural characteristics, optical properties, and electrical conductivity of doped [P(An-MMa)/ZrO2]TF nanofiber composite using experimental data and TD-DFT/DMol3 computations, *Environmental Science and Pollution Research,* (2022) 1-19.

[39] Kenawy, E. R., Samar Abdelhady, Mohamed M Azaam, Chemical modification, electrospinning and biological activities of pluronic F68, *Polymer Bulletin,* (2022) 1-16.

[40] Azaam, M., El-Refaie Kenawy, Abd El Hay AM, Nermeen Saad, Synthesis and characterization of poly L-Lactic acid by direct polycondensation, *Delta Journal of Science* (2022) 44 (2), 155-165.

[41] Tamer, T. M., E. R. Kenawy, M. M. Agwa, S. A. Sabra, M. A. El-Meligy, M. S. Mohy-Eldin, Wound dressing membranes based on immobilized Anisaldehyde onto (chitosan-GA-gelatin) copolymer: *In-vitro* and *in-vivo* evaluations, *International Journal of Biological Macromolecules* (2022) 211, 94-106.

[42] Kenawy, E. R., Elbadawy A Kamoun, Mahmoud Serag Eldin, Hesham Soliman, Shahira H EL-Moslamy, Esmail M El-Fakharany, Abdel-baset M Shokr, Electrospun PVA–Dextran Nanofibrous Scaffolds for Acceleration of Topical Wound Healing: Nanofiber Optimization, Characterization and *In Vitro* Assessment, *Arabian Journal for Science and Engineering*, (2022) 1-18.

[43] Khater, S. F., R Ghazy, N Hendawya, El-Refaie Kenawy, Ahmed Farouk Elmekawy, M H Makled, Ahmed R Ghazy, Experimental static light scattering study on the scattering parameters for Pluronic, *Benha Journal of Applied Sciences* (2022) 7 (5), 257-263.

[44] Kenawy, E. R., Azza A Shabaka, Alaa M Abou-Zeid, Mohamed S Hassouna, Mohamed A Elhiti, Effective biological treatment of water polluted with coomassie brilliant blue and methylene blue using carbon nanotube-supported biodegradation, *Environmental Progress & Sustainable Energy* (2022) 41 (3), e13783.

[45] Kenawy, E. R., Heikki Tenhu, Samar A Khattab, Ahmed A Eldeeb, Mohamed M Azaam, Highly efficient adsorbent material for removal of methylene blue dye based on functionalized polyacrylonitrile, *European Polymer Journal* (2022) 169, 111138.

[46] El-Gharably, A. A., El-Refaie S Kenawy, Ahmed A Safaan, Saad A Aboamna, Yehia A-G Mahmoud, Hamada SA Mandour, Synthesis, characterization and application of chitosan conjugated heterocyclic compounds, *Journal of Polymer Research* (2022) 29 (4), 1-10.

[47] Alothman, A. A., Mehrez E El-Naggar, M Afifi, Mohammed Sheikh Saleh Mushab, Mika Sillanpää, El-Refaie Kenawy, Optimizing Graphene Oxide Encapsulated TiO2 and Hydroxyapatite; Structure and Biological Response, *Journal of Inorganic and Organometallic Polymers and Materials* (2022) 32 (4), 1306-1318.

[48] Kenawy, E. R., Mohamed Labib Salem, Noura Sanoh, Mohamed Azaam, T1107 polymer expressed potential anti-tumor effects against HepG2 cell line, *International Journal of Cancer and Biomedical Research* (2022) 6 (1), 1-10.

[49] Elmaghraby, N. A., Ahmed M Omer, El-Refaie Kenawy, Mohamed Gaber, Ahmed El Nemr, Electrospun composites nanofibers from cellulose acetate/carbon black as efficient adsorbents for heavy and light machine oil from aquatic environment, *Journal of the Iranian Chemical Society,* (2022) 1-15.

[50] El-Shanshory, A. A., Mona M Agwa, Ahmed I Abd-Elhamid, Hesham MA Soliman, Xiumei Mo, El-Refaie Kenawy, Metronidazole topically immobilized electrospun nanofibrous scaffold: Novel secondary intention wound healing accelerator, *Polymers* (2022) 14 (3), 454.

[51] Moharir, K. S., Mallesh Kurakula, Vinita Kale, El-Refaie Kenawy, Sheikh Murtuja, Mohd Neyaz Ahsan, Md Saquib Hasnain, Amit Kumar Nayak, Biomedical applications of electrospun chitosan nanofibers, *Chitosan in Biomedical Applications*, (2022) 75-110.

[52] Kenawy, E. R., Azza A. Shabaka, Allaa M. Abou-Zeid, Mohamed S. Hassouna, Mohamed A. Elhiti, Silica-based nano-adsorbent for enhancement of bacterial biodegradation of methylene blue dye, *Desalination and Water Treatment,* (2021) 1–9, doi: 10.5004/dwt.2021. 27829.

[53] Kenawy, E. R., Aza Shabaka, Alaa Abou Zeid, Effective biological treatment of water polluted with coomassie brilliant blue and methylene blue using carbon nanotube-supported biodegradation, *Environ Prog Sustainable Energy*. 2021; e13783. Corrected Proof.

[54] El-Gharably, A. A., El-Refaie S. Kenawy, Ahmed A. Safaan, Saad A. Aboamna, Yehia A. G. Mahmoud, Synthesis, characterization, antimicrobial and anticancer evaluation of *N*-aryl amino chitosan, *Polymer Bulletin* https://doi.org/10.1007/s00289-021-03960-y.

[55] Ansari, M. T., Md Saquib Hasnain, Amit Kumar Nayak and El-Refaie Kenawy, book chapter, Chitosan-based nanobiocomposites in drug delivery 411 published in *Chitosan in Drug Delivery*, Edited by Md Saquib Hasnain, Sarwar Beg, Amit Kumar Nayak, Academic Press. ISBN: 978-0-12-819336-5, 2021. Link Here.

[56] Saleh, A. Mohamed Nesma M. Elaraby Azza M. Abdel-Aty Elkhabiry Shaban, M. A. Abu-Saied, El-Refaie Kenawy, Mehrez E. El-Naggar "Improvement of enzymatic properties and decolorization of azo dye: Immobilization of horseradish peroxidase on cationic maize starch," *Biocatalysis and Agricultural Biotechnology, Available online* 11 November 2021,102208. Link Here.

[57] Mehrez, E. El-Naggar, Ola A. Abu Ali, Dalia I. Saleh, M. A. Abu-Saied, M. K. Ahmed E. Abdel-Fattah, S. F. Mansourm El-Refaie Kenawy, Facile modification of polycaprolactone nanofibers with hydroxyapatite doped with thallium ions for wound and mucosal healing applications, *Journal of Materials Research and Technology* In press, Available online 24 September 2021.

[58] Kenawy, E. R., Maurizia Seggiani AliHosny· Mohamed Rashad· Patrizia Cinelli Khalil M. Saad-Allah Mohamed El-Sharnouby S. Shendy Mohamed M. Azaam Superabsorbent composites based on rice husk for agricultural applications: Swelling behavior, biodegradability in soil and drought alleviation, *Journal of Saudi Chemical Society*, Volume 25, Issue 6, June 2021, 101254.

[59] Elbadawy, A. Kamoun, Samah A. Loutfy, Yasmein Hussein, El-Refaie S. Kenawy, Recent advances in PVA-polysaccharide based hydrogels and electrospun nanofibers in biomedical applications: A review, *International Journal of Biological Macromolecules Volume 187*, 30 September 2021, Pages 755-768. https://ejabf.journals.ekb.eg/?_action =current&_is=Current%20Issue.

[60] Kenawy, E. R., Ali Hosny, Khalil Saad-Allah, Reducing nitrogen leaching while enhancing growth, yield performance and physiological traits of rice by the application of controlled-release urea fertilizer *Paddy and Water Environment* volume 19, pages 173–188 (2021).

[61] Nemr, A. E., Aida H. Shobier, El Sayed H. El Ashry, Sherif H. Kandil, El-Refaie Kenawy, Pharmacological screening of some Egyptian marine algae and encapsulation of their bioactive extracts in calcium alginate beads, *Egyptian Journal of Aquatic Biology & Fisheries,* Vol. 25(2): 1025-1052 (2021).

[62] Elbarbary, A. A., El-Refaie Kenawy, Elsayed G I Hamada, Tarek B Edries, Wesam S Meshrif, Insecticidal activity of some synthesized 1,3,4-oxadiazole derivatives grafted on chitosan and polymethylmethacrylate against the cotton leafworm *Spodoptera littoralis Full text links Int J Biol Macromol*. 2021 Mar 17;180:539-546. doi:10.1016/j.ijbiomac.2021.03.050.

[63] Grad, W. E., M. M. A. Ghallab, E. R. Kenawy, M. I. Massoud Impact of rice husk polymer composite application under deficit irrigation on productivity and stress tolerance of some sugarcane varieties, *Direct Research Journal of Agriculture and Food Science*, Vol. 9, Pp. 218-229, 2021.

[64] Elmessiry, K. S., M. R. El-Aassar, A. B. A. A Nassr, El-Refaie Kenawy, Baha Eddin Moharam & Nehal Ali. Free-Standing Working Electrodes for Supercapacitors Based on Composite Polymer Nanofibers and Functionalized with Graphene Oxide. *J. Electron. Mater.* (2021). https://doi.org/10.1007/s11664-021-09120-2.

[65] El-Refaie Kenawy, Mohamed Azam, Syed Aneesh Ahmed and Md Saquib Hasnain, *Antimicrobial Nanocomposite, in Biomedical composites*, Amit Kumar Nayak, Md Hasnain Editors, Springer 2021 page 71.

[66] Abdel-Karim, A., Mehrez E. El-Naggar, E. K. Radwan, Ibrahim M. Mohamed Mohamed Azaam, El-Refaie Kenawy, High-performance mixed-matrix membranes enabled by organically/inorganic modified montmorillonite for the treatment of hazardous textile wastewater, *Chemical Engineering Journal* Volume 405, 1 February 2021, 126964.

[67] Eltorky, M., Mahmoud shakal; Waleed elshahawy; El-Refaie Kenawy, Effect of Simulated Gastric Acid on Surface Roughness of Different Types of Dental Ceramics, *Egyptian Dental Journal Article* 22, Volume 67, Issue 1 - January (Fixed Prosthodontics, Removable Prosthodontics and Dental Materials), January 2021, Page 711-716.

[68] Sameh, M. Osman, Fatmah A. Alasmary, El-Refaie Kenawy, El-Saied A. Aly, Sherine N. Khattab & Ayman El-Faham, Synthesis, characterization and comparative thermal degradation kinetics of *s*-Triazine based polymers, *Journal of Polymer Research* volume 28, Article number: 304 (2021), https://doi.org/10.1007/s10965-021-02667-yhttps://link.springer.com/article/10.1007/s10965-021-0266, https://rdcu.be/cC2y4.

[69] Rashad, M., El-Refaie Kenawy, Ali Hosny, Mohamed Hafez, and Maha Elbana An environmentally friendly superabsorbent composite based on rice husk as soil amendment to improve plant growth and water productivity under deficit irrigation conditions, *Journal of Plant Nutrition* 2021, Vol. 44, No. https://www.tandfonline.com/doi/abs/10.1080/01904167.2020.1849293?journalCode=lpla20.

[70] Kenawy, E. R., Mohamed M. Azaam, Mohd Afzal, Aisha Khatoon, Mohammed Tahir Ansari, Md Saquib Hasnain, *Pharmaceutical application of cellulose derivatives, Tailor-Made Polysaccharides in Biomedical Applications,* AK Nayak, MS Hasnain, TM Aminabhavi Editors- Academic Press 2021.

Mohamed M. Azaam

Mohamed M. Azaam has been a Professor in the Chemistry Department, Faculty of Science, Tanta University, since 2023. He supervises 10 master's students and 4 PhD students and has published approximately 41 manuscripts in international journals along with 3 book chapters. In 2018, he received an Encouragement Award from Tanta University.]

Chapter 6

Integrating Biostimulants with Irrigation Management to Promote Sustainable Agriculture

Qurat-Ul-Ain Raza[1,2]
Abdur Rehim[1]
Asad Ismail Noor[1]
Ijaz Ahmad[3]
Faiz-Ul-Hassan[4]
and Muhammad Amjad Bashir[5,*]

[1]Department of Soil Science, Faculty of Agricultural Sciences and Technology, Bahauddin Zakariya University, Multan, Pakistan
[2] Department of Biological and Environmental Sciences, Emerson University Multan, Multan, Pakistan
[3]Department of Agronomy, Engro Fertilizers Ltd., Lahore, Pakistan
[4]Department of Soil Science, College of Agriculture, University of Layyah, Layyah, Pakistan
[5]Sichuan Provincial Key Laboratory of Philosophy and Social Sciences for Monitoring and Evaluation of Rural Land Utilization, Chengdu Normal University, Chengdu, P. R. China

Abstract

The increasing world population has forced the excessive use of mineral fertilizer and irrigation to complete global food demand, deteriorating the water quality and nutrient losses. Biostimulants have a promising role in irrigation management as well as achieving sustainable agricultural

* Corresponding Author's Email: hhan@siumed.edu

In: Irrigation Management
Editors: Maha Elbana and Youssef Brouziyne
ISBN: 979-8-89530-318-4

production. The application of biostimulants has the potential to support plants to survive under water-deficient conditions, improve crop production, minimize the application of mineral fertilizers leading to reduced water pollution, and enhanced nutrient use efficiency. In this chapter, we have focused on the role of biostimulants in irrigation management, water deficit conditions, and water pollution to promote sustainable agriculture. The application of biostimulants is an emerging technology that can contribute to plant growth and irrigation management. In addition, the challenges and future aspects are also addressed. To ensure the quality of the chapter, we collected references from well-reputed international journals using the keywords "biostimulants," "drought," "water quality," "water pollution," and "sustainable agriculture." In conclusion, biostimulants can contribute to irrigation management, reducing water pollution, and sustainable agriculture.

Keywords: drought, leeching, water pollution, water quality, water stress

List of Abbreviations

ATP	Adenosine triphosphate
AMP	Arbuscular mycorrhizal fungi
PGPR	Plant growth Promoting Rhizobacteria
TBA	Terbuthylazine

Introduction

The world population has been estimated to reach 10 billion by the end of 2050 and this huge population requires a 70% increase in food demand (Raza et al. 2022). Chemical fertilizers are usually applied to target the huge food production goal. It is estimated that about 1.9×10^{11} tons of chemical fertilizers are used globally (Wan et al. 2021) but the excessive use of chemical fertilizers results in nutrient loss such as leeching, thus deteriorating ground and surface water (Cui et al. 2020).

Studies also reported that the crops uptake only 30-50% of the nitrogen fertilizers whereas the remaining is lost in the soil and pollutes the groundwater (Liu et al. 2023; Tyagi et al. 2022). For the last two decades, the application of nitrogen fertilizers has increased but the nitrogen use efficiency has declined (Wang et al. 2018). In addition, chemical fertilizers also decline soil fertility status, and soil eco-profile; decline soil organic matter, and disturb biodiversity and biological activity (Pahalvi et al. 2021).

Water is essential for plant growth and a large amount of water is consumed during cultivation. Using excessive irrigation during crop cultivation also results in nutrient losses and water quality deterioration (Wang et al. 2020). Besides this, global warming and climatic water balance have been reported to increase adversely and result in water shortage (Bisbis et al. 2018).

Drought is also an environmental hazard associated with the shortage of water. This water stress disturbs the plant performance, enzymatic activities, and nutrient uptake, and adversely affects plant production (Bashir et al. 2022). Moreover, it is also associated with severe financial hardship, to malnutrition, and even famine in rich and poor countries respectively (Passioura 2007). It also affects the social, economic, and the productivity and functioning of soil systems (Geng et al. 2015). In this scenario, food security along with sustainable agriculture and irrigation management is a major challenge to ensure huge crop production and water quality for the agricultural scientists and communities.

Biostimulants are organic or inorganic substances that when applied to plants improve its ability to develop resistance against biotic or abiotic stresses, improve nutrient use efficiency, and enhance the quality of food products (Bashir et al. 2022). Various benefits of biostimulants have been reported and are gaining popularity throughout the world. In the absence of chemical fertilizers, biostimulants when applied to radish improved its photosynthetic activity, chlorophyll content, and yield (Raza et al. 2022). In addition, it also induce positive changed in numerous crops including turnip (Raza et al. 2023), tomato (Alfosea-Simón et al. 2021), radish (Rehim et al. 2021), and yellow pepper (Paradiković et al. 2011).

Besides this, it has also been reported to develop resistance in plants against water stress. Exogenous application of amino acids enhanced the growth and nutritional value of cabbage under water stress (Haghighi et al. 2020). Moreover, glycine betaine can be proposed as a viable biostimulants to cultivate maize in a water-deprivation regimen (Jiménez-Arias et al. 2022). Potentially toxic elements produced by the industries are also a cause behind contamination in arable soil and it also pollutes aquatic and terrestrial environments as well as humans. Biostimulants also have the potential to serve in this respect. A study reported that foliar application of Zinc-Iron-Lysine improved rapeseed (Brassica napus L.) growth and composition even when irrigated with tannery wastewater (Zaheer et al. 2022).

Besides this, biostimulants have also been introduced as a tool that contributes to the bioconversion compost of organic waste. It is an innovative organic waste management technology that contributes to environmental safety and sustainable agricultural development (Huang et al. 2021). Moreover, it also reduces microbial activity during composting and ultimately minimizes greenhouse gas emissions (Ermolaev et al. 2015). Some limitations of biostimulants include crop specificity of biostimulants, timing considerations, and environmental factors (Rehim et al. 2021). Considering all such benefits of biostimulants, this chapter aims to elaborate on the role of biostimulants in sustainable agriculture and irrigation management. To ensure the quality of the chapter, we collected references from well-reputed international journals using the keywords "biostimulants," "drought," "water quality," "water pollution," and "sustainable agriculture." This chapter focuses on the role of biostimulants in irrigation management, water deficit conditions, and water pollution to promote sustainable agriculture. The chapter consists of 6 sections, which describe a comprehensive overview of biostimulants and the pressing need for their integration into agricultural practices (Section 1), elucidate the intricate interplay between water management strategies, and highlight its significance (Section 2), explore the role of biostimulants to bolster crop resilience in water-limited environments (Section 3), shed light on the potential of biostimulants as eco-friendly solutions to environmental contamination specifically water pollution (Section 4), navigate the complexities of biostimulants application and associated challenges as well as future directions (Section 5), and lastly the key insights from our exploration and recommendations (Section 6), respectively.

Precise Irrigation and Its Significance

Due to the increasing population, the water demand has also been projected to increase significantly in the coming years. Over the last century, the irrigated area of the world has increased more than 6 folds and has reached 260 million hectares and is increasing 1% each year (Chartzoulakis and Bertaki 2015). In addition, the agriculture sector is the largest consumer with usage of 65-75% of freshwater as irrigation which may increase up to 90% in some cases (de Fraiture et al. 2010). However, to ensure food security challenges, it is crucial to sustain irrigation water and avoid wastage of water.

Herein, it is important to mention that overuse of irrigation water also depletes groundwater quality as well as impacts food security (Zhang et al. 2023). Maize crop yield is influenced by the soil water content. This result was similar to the farmer's technique in that sufficient water application (irrigation) at the start of the crop season saves the water in the soil to be used for the time when crop water requirement is high. This concept is named "soil-water banking" in Colorado (de Lara et al. 2019).

Precise irrigation can be defined as the accurate determination and quantification of the water requirement of the crop and its precise application at the required time (Adeyemi et al. 2017). Therefore, precise irrigation is a useful technique to enhance crop production and supply water to the desirable location considering the crop requirement. It is used as a tool to achieve sustainability in irrigation, water quality, and water use efficiency (Abioye et al. 2020). A study also reported that precise irrigation technique has the potential to improve average water saving potential as well as yield potential in both rainy and dry years (Souza et al. 2023). In addition, water saving can be increased from 9-19% with precision irrigation compared to conventional irrigation practices (Hedley and Yule 2009). A case study reported that a variable rate irrigation strategy improved the crop yield as well as reduced the percolation losses specifically during the dry season (González Perea et al. 2018). Another three-year study conducted in China demonstrated that precise irrigation and planting patterns have the potential to develop suitable growing conditions for wheat and also increase grain yield (Mao et al. 2017).

Numerous irrigation management strategies have introduced various tools and instruments that save water as well as maintain crop growth and quality. For instance, soil moisture sensor controllers (20–92%), evapotranspiration controllers (20–71%), and rain sensors (7–50%) have contributed to saving water (Touil et al. 2022). In addition, optical sensors (aerial vehicles and drones) and the Internet of Things (IoT) have also been used for irrigation management (Touil et al. 2022). Precise irrigation also has some adoption limitations such as initial cost (Ara et al. 2021), technical expertise of the labor (Hertel and Lobell 2014), and climatic conditions (Ward 2022).

Sustainable water management is associated with multiple agricultural practices including soil management to stabilize soil resources and reduce erosion, fertilizer application according to crop requirements, pest and disease control by ensuring sustainable practices, and irrigation management to meet crop needs and minimize water losses (Figure 1). The utilization of agrochemical inputs, fertilizers, and pesticides contributes to water quality deterioration as well as adversely affects the environment (Mitra et al. 2022).

Integrating precise irrigation practices and water management techniques can also reduce surface runoff losses and soil erosion, improve the environmental quality of irrigated fields, improve groundwater quality, and accurate estimation of soil moisture content can also help in irrigation scheduling decisions (Adeyemi et al. 2017). Various technologies have been introduced to achieve sustainable water management in agriculture. One of the latest is biostimulants that we will discuss further in two aspects i.e., water stress and water pollution. Biostimulants are widely being used to sustain plant growth under water stress conditions and are being introduced as a tool to reduce water pollution.

Role of Biostimulants Against Water Stress

Plants encounter a wide range of environmental problems that have a detrimental influence on their growth, metabolism, and productivity. Drought is the principal abiotic factor limiting agricultural crop productivity, and plant responses to water stress differ depending on development stage, as well as the severity and duration of the stress (Reddy et al. 2004). Drought is often recognized as the greatest significant threat to farmers growing field crops, and its frequency and severity have increased worldwide (He et al. 2022). Drought is a constant, recurring aspect of climate that occurs in practically all climatic regimes; even in more humid climatic zones, drought is commonly present. It also has a significant influence on agriculture (Goñi et al. 2018) as it impacts on agricultural productivity and output have major implications for food security and the economies of the affected regions (Carrão et al. 2016). The frequency of agricultural drought is governed by crop evapotranspiration demand and soil moisture availability to meet this requirement (White 2010). Water stress can cause photosynthetic problems, metabolic disturbance, water content reduction, decreased cell growth and expansion, and even plant death (Jaleel et al. 2008). Nutrient deficits can occur due to dry soil conditions (Afshar et al. 2014).

In water-scarce settings, biostimulants can be used as an alternative to boost crop yield by protecting plants from environmental stresses and reducing their negative effects (Calvo et al. 2014). Many crops have embraced the foliar or soil application of biostimulants to enhance plant physiology and metabolism (Mariani and Ferrante 2017). Biostimulants can have a wide range of specific or broad action mechanisms (Łangowski et al. 2019). In the current context of rapidly expanding climate change, microbial biostimulants provide a feasible alternative for sustaining plants subjected to abiotic stresses (Santoyo et al. 2021). Ascophyllum nodosum seaweed extract is used in biostimulants, which encourage vegetative growth, increase nutrient absorption efficiency, boost antioxidant responses, and accumulate cellular defense metabolites. These properties provide plants with increased resistance to environmental stressors and yield losses (Jithesh et al. 2019).

Figure 1. Major aspects of sustainable water management.

The morphological and metabolic changes that plants undergo in response to orthosilicic acid and seaweed extract help to mitigate the damaging effects of drought. Moreover, it enhance antioxidant and photoprotective defense mechanisms while also increasing water usage efficiency (Sujata et al. 2023). Fulvic acid serves to enhance respiration, photosynthesis, and chlorophyll contents, as well as to promote the growth of roots, leaves, and shoots. It also enhances ion absorption and ATP generation (Calvo et al. 2014). Microbial biostimulants, notably arbuscular mycorrhizal fungi (AMF) and plant growth-promoting rhizobacteria (PGPR), have the potential to be useful long-term treatments, in contrast to the creation of various stress-tolerant types. These environmentally friendly instruments can help manage drastically climate-resilient farmlands and enhance the health condition of plants under abiotic stress (Ngoroyemoto et al. 2020).

Table 1. Response of plants against water stress using biostimulants

Plant	Biostimulant	Plant response	References
Tomato	Pepton	Exerts a favourable influence on tomato leaves' hormonal profile and improves abiotic defences under water stress circumstances, including defense-related phytohormones and antioxidants.	(Casadesús et al. 2019)
Mint	CRADLE™, Mobilizer™, Nanozim De'Lite™, Nanozim NXT™	Plants treated with biostimulants showed improved morphological traits, physiological performance, and EO ratio.	(Elansary et al. 2020)
Tomato	ERANTHIS®®	The use of biostimulants reduces stress in tomato plants.	(Campobenedetto et al. 2021)
Lettuce	Codasil®	Codasil® promoted plant growth under water-stressed situations, decreased lipid peroxidation and H2O2 levels, and improved photosynthetic performance.	(Hidalgo-Santiago et al. 2021)
Green Bean	Nomoren, EKOprop, Veramin Ca, Twin-Antistress	The use of biostimulants improved the nutritional content and chemical composition of pods and seeds.	(Petropoulos et al. 2020)
Wheat	Humic Acid (HA)	It was demonstrated that using HA from the leaves to alleviate drought stress in natural settings will be helpful in stress-tolerant wheat farming.	(Arslan et al. 2016)
Petunia	LME (Moringa oleifera leaf extract)	Leaf moringa treatment boosted plant drought tolerance, particularly by reducing the impacts of water deficiency on growth metrics, proline and MDA levels, and enzyme activity.	(Toscano et al. 2023)
Eucalyptus urophylla	Stimulate®	Stimulate® may be a feasible solution for mitigating the physiological consequences of negative water stress in young Eucalyptus urophylla plants.	(Santos et al. 2019)
Olive	Moringa leaves aqueous extract (MLE 5%)	MLE, an eco-friendly and affordable treatment, may be used to increase the yield of olive trees in water deficiency situations.	(Abidi et al. 2023)
Thale cress	Delfan Plus Phylgreen Trafos K	Each of the biostimulants examined improved plant growth and reduced the impacts of drought stress. Although each of the three biostimulants increased growth, they functioned differently depending on the application rate and time of application.	(Fleming et al. 2019)

Plant	Biostimulant	Plant response	References
Pepper	Concentrated Marine Algae Extracts (CMAEs)	The application of 0.5% CMAEs significantly increased the stomatal conductance of pepper plants. The lowest dose of this product demonstrated greater efficiency in all physical and chemical parameters tested.	(Melo et al. 2020)
Faba Bean	Fucus spiralis, Ulva lactuca, Laminaria ochroleuca, Ascophyllum nodosum	Seaweed extracts can reduce drought stress via several processes including a number of bioactive chemicals.	(El Boukhari et al. 2023)
Sesame	Salicylic Acid, Beeswax Waste Extract, Licorice Extract (LE)	Exogenous treatment of biostimulants during stress developed drought resistance in plants.	(Pourghasemian et al. 2020)
Soy bean	*Ascophyllum nodosum*	Conserve chlorophyll content, improve photosynthesis and help plant develop resistance against water stress.	(Melo et al. 2024)
Cucumis melo L.	Aqueous extracts (*Ilex paraguariensis* A.St.-Hil.)	Activate the enzymatic and non-enzymatic antioxidant defense mechanisms, reduce oxidative stress, and mitigate the water scarcity in melon crops.	(Toscano Adamo et al. 2024)
Maize	Natural extracts from medicinal plants including Moringa oleifera leaf, bark, Aegel marmelos leaf, Phyllanthus niruri leaf, Terminalia arjuna leaf and bark.	Improved the antioxidant activity, physiological and biochemical performance, enhanced the seedling establishment and growth under water deficient conditions.	(Bhowmick et al. 2024)
Vitis vinifera L.	Bacterial-based formulations	Boosted the plant metabolism and developed resistance against drought stress.	(Papantzikos et al. 2024)

Several studies reported that biostimulants have the potential to develop tolerance in plants against water stress (Table 1). Biostimulants are a source of protein, amino acids, and primary and secondary metabolites that enhance the scavenging effectiveness, and reactive oxygen species, and promote the antioxidant defense mechanism in plants against oxidative stress (Abd El-Mageed et al. 2017). It also improves stomatal regulation and ultimately photosynthesis, plant growth (Baghdadi et al. 2022), and root development (Rehim et al. 2024). Moreover, soil application of biostimulants also influences the microbial as well as nutrient dynamics in soil which also increases crop growth (Chen et al. 2002).

Integration of biostimulants with different irrigation regimes also improved almond production and yield specifically under sustained-deficit irrigation (Gutiérrez-Gordillo et al. 2019). However, a combination of biostimulants with different irrigation methods needs to be studied along with water quality to explore water and agricultural sustainability. Additionally, the response of plants toward biostimulants depends upon the formulation, doses, timing, climatic conditions, and crop species. Multiple trials and research experiments need to be planned considering these research gaps for optimization of biostimulants in agriculture.

Role of Biostimulants Against Water Pollution

Irrigation plays a significant role in agricultural intensification and meeting food security challenges. But irrigation and drainage are also associated with water quality deterioration probably due to runoff and leaching of fertilizers, pesticides, and salts. Po Plain of Northern Italy is a good example of nitrate pollution (Rotiroti et al. 2023). Nearly 50% of the N fertilizer is used by plants while remaining is lost as evaporation (2-20%), evaporation (15-25%), surface and ground water (2-10%). Phosphorus fertilizers also contribute towards eutrophication, and their long-term fertilization in paddy reported an increased TP in surface water. It also contributes towards non-point source pollution and increase the risk of surface and ground water pollution (Khan et al. 2017). The consumption of chemical fertilizers has increased ten times since the 1960s (Mateo-Sagasta et al. 2017). As discussed earlier, excessive use of chemical fertilizers contributes to water pollution, eutrophication, and deterioration of surface and groundwater. So, it is crucial to identify the role of biostimulants in plant growth, fertilization management, and its significance towards sustainable agricultural production. Recently studies reported that biostimulants can be used to minimize or even substitute chemical fertilizers to increase vegetable production (Raza et al. 2022). In addition, vitamin B12 and coenzyme Q10 when applied to plants in the absence of chemical fertilizers improved the radish yield under controlled conditions (Rehim et al. 2021). This is a clear identification that biostimulants may contribute to reducing water pollution due to chemical fertilizers.

Apart from chemical fertilizers, pesticides, and herbicides also contribute to exacerbating water pollution by replacing or altering with biostimulants. A study reported that duckweed when integrated with biostimulants is suitable for cleaning water polluted with terbuthylazine (TBA) (Panfili et al. 2019). Besides this, biostimulants have the potential to alleviate metal stress and ensure plant stability when irrigated with tannery wastewater (Zaheer et al. 2022). Biostimulants can contribute towards sustainable agriculture and contribute towards sustainable agriculture, mitigating the use of chemical fertilizers, reducing pesticide pollution, and ensuring plant growth under polluted water irrigation. But the direct role in controlling water pollution isn't well reported.

Challenges and Future Directions

Biostimulants have been reported to support plant growth under water stress conditions but long-term studies under natural conditions should be addressed. In addition, biostimulants also support plant growth under contaminated water irrigation but only a few studies have been reported to treat polluted water therefore this aspect can be studied in the future. Short-term pot studies revealed that foliar application of biostimulants has the potential to replace chemical fertilizers. Moreover, drought or water stress is also a major aspect associated with irrigation. Biostimulants have the potential to develop tolerance in plants against water stress. Depending upon the species and climatic conditions, biostimulants can be selected and applied to ensure plant growth and development under water stress.

Furthermore, water-saving technologies such as precision irrigation, drip fertigation, and soil moisture management techniques can be integrated with the use of biostimulants to address the issue of N losses and eutrophication leading to water pollution. Biostimulants have now also been introduced as phytoremediation tools because of their capability to sustain plant growth against toxic environmental conditions as well as remediate the contaminated environment. Although all biostimulants have not been tested as phytoremediation further studies are needed. Moreover, greenhouse gas emissions and the role of biostimulants compared to common agricultural practices are also a research gap. Due to the uniqueness of biostimulants, the mechanism of action of biostimulants is revealed yet. However, the identification of the role of biostimulants in field experiments and its long-term influence on crop production and environmental sustainability need to be studied. Furthermore, multi-disciplinary approaches such as agronomic, physiological, and molecular analyses to understand the influence of biostimulants and their interaction with plants and soil microbiota under varying environmental conditions can contribute to identifying the potential mechanism. Additionally, biostimulants are usually considered expensive but detailed studies in ambient conditions should be performed to indicate their economic potential compared to chemical fertilizers. In addition, life-cycle assessment can clarify the environmental footprint of agricultural inputs specifically linked with precise irrigation techniques combined with biostimulants. Nutrient leaching and runoff losses due to chemical fertilizers should also be compared with biostimulants to identify their role in water pollution and environmental sustainability. Thus, further studies need to be conducted to identify more benefits of biostimulants specifically towards sustainable agriculture, and irrigation management.

Conclusion

With the increase in world population, food demand has also increased huge fertilization consumption and water pollution. In addition, climate change also led to drought conditions creating unfavorable conditions for crops. Biostimulants have a promising role in irrigation management and water stress. Few studies reported that biostimulants have the potential to deal with multiple challenges in agriculture such as water scarcity, environmental pollution, and nutrient management, and promote sustainable crop production. The chapter summarizes that biostimulants are also capable of mitigating the use of chemical fertilizers as well as sustaining plant growth under stress conditions and fertilization-reduced environments. Integration of biostimulants with water management practices can also be explored to reach sustainable agriculture and environment. The crop species and the climatic conditions might differ in response to the application of biostimulants. Therefore, field experiments and long-term trials might help indicate the mechanism behind the efficacy of biostimulants in agriculture. In conclusion, biostimulants integrated with irrigation management practices might be a solution to sustainable agriculture.

Disclaimer

None

References

Abd El-Mageed TA, Semida WM, Rady MM (2017) Moringa leaf extract as biostimulant improves water use efficiency, physio-biochemical attributes of squash plants under deficit irrigation. *Agric Water Manag* 193:46–54. https://doi.org/10.1016/j.agwat.2017.08.004.

Abidi W, Akrimi R, Gouiaa M (2023) Mitigating drought stress by plant-derived biostimulant in Arbequina olive (Olea europeae L.) cultivar conducted in super high density. *Acta Physiol Plant* 45:1–17. https://doi.org/10.1007/s11738-023-03613-9.

Abioye EA, Abidin MSZ, Mahmud MSA, Buyamin S, Ishak MHI, Rahman MKIA, Otuoze AO, Onotu P, Ramli MSA (2020) A review on monitoring and advanced control strategies for precision irrigation. *Comput. Electron. Agric.* 173:105441.

Adeyemi O, Grove I, Peets S, Norton T (2017) Advanced monitoring and management systems for improving sustainability in precision irrigation. *Sustain.* 9:353.

Afshar RK, Chaichi MR, Assareh MH, Hashemi M, Liaghat A (2014) Interactive effect of deficit irrigation and soil organic amendments on seed yield and flavonolignan production of milk thistle (Silybum marianum L. Gaertn.). *Ind Crops Prod* 58:156–172. https://doi.org/10.1016/j.indcrop.2014.03.043.

Alfosea-Simón M, Simón-Grao S, Zavala-Gonzalez EA, Cámara-Zapata JM, Simón I, Martínez-Nicolás JJ, Lidón V, García-Sánchez F (2021) Physiological, Nutritional and Metabolomic Responses of Tomato Plants After the Foliar Application of Amino Acids Aspartic Acid, Glutamic Acid and Alanine. *Front Plant Sci* 11:581234. https://doi.org/10.3389/fpls.2020.581234.

Ara I, Turner L, Harrison MT, Monjardino M, DeVoil P, Rodriguez D (2021) Application, adoption and opportunities for improving decision support systems in irrigated agriculture: A review. *Agric. Water Manag.* 257:107161.

Arslan E, Agar G, Aydin M (2016) Humic Acid as a Biostimulant in Improving Drought Tolerance in Wheat: the Expression Patterns of Drought-Related Genes. *Plant Mol Biol Report* 39:. https://doi.org/10.1007/s11105-020-01266-3.

Baghdadi A, Della Lucia MC, Borella M, Bertoldo G, Ravi S, Zegada-Lizarazu W, Chiodi C, Pagani E, Hermans C, Stevanato P, Nardi S, Monti A, Mangione F (2022) A dual-omics approach for profiling plant responses to biostimulant applications under controlled and field conditions. *Front Plant Sci* 13:983772. https://doi.org/10.3389/fpls.2022.983772.

Bashir MA, Raza Q-U-A, Rehim A, Umair Sial M, Muhammad Ali Raza H, Ali S, Ijaz M, Faiz-Ul-Hassan, Geng Y (2022) Biostimulants Application: An Innovative Approach to Food Security under Drought Stress. In: *Climate Change - Recent Observations* [Working Title]. IntechOpen.

Bhowmick S, Rai G, Mishra SK, Bisht N, Chauhan PS (2024) Bio-stimulants from medicinally and nutritionally significant plant extracts mitigate drought adversities in Zea mays through enhanced physiological, biochemical, and antioxidant activities. *Plant Physiol Biochem* 207:108396. https://doi.org/10.1016/j.plaphy.2024.108396.

Bisbis MB, Gruda N, Blanke M (2018) Potential impacts of climate change on vegetable production and product quality – A review. *J. Clean. Prod.* 170:1602–1620.

Calvo P, Nelson L, Kloepper JW (2014) Agricultural uses of plant biostimulants. *Plant Soil* 383:3–41. https://doi.org/10.1007/s11104-014-2131-8.

Campobenedetto C, Agliassa C, Mannino G, Vigliante I, Contartese V, Secchi F, Bertea CM (2021) A Biostimulant Based on Seaweed (Ascophyllum nodosum and Laminaria digitata) and Yeast Extracts Mitigates Water Stress Effects on Tomato (Solanum lycopersicum L.). *Agric 2021*, Vol 11, Page 557 11:557. https://doi.org/10.3390/AGRICULTURE11060557.

Carrão H, Naumann G, Barbosa P (2016) Mapping global patterns of drought risk: An empirical framework based on sub-national estimates of hazard, exposure and vulnerability. *Glob Environ Chang* 39:. https://doi.org/10.1016/j.gloenvcha.2016.04.012.

Casadesús A, Polo J, Munné-Bosch S (2019) Hormonal effects of an enzymatically hydrolyzed animal protein-based biostimulant (pepton) in water-stressed tomato plants. *Front Plant Sci* 10:. https://doi.org/10.3389/fpls.2019.00758.

Chartzoulakis K, Bertaki M (2015) Sustainable Water Management in Agriculture under Climate Change. *Agric Agric Sci Procedia* 4:88–98. https://doi.org/10.1016/j.aaspro.2015.03.011.

Chen SK, Subler S, Edwards CA (2002) Effects of agricultural biostimulants on soil microbial activity and nitrogen dynamics. *Appl Soil Ecol* 19:249–259. https://doi.org/10.1016/S0929-1393(02)00002-1.

Cui N, Cai M, Zhang X, Abdelhafez AA, Zhou L, Sun H, Chen G, Zou G, Zhou S (2020) Runoff loss of nitrogen and phosphorus from a rice paddy field in the east of China: Effects of long-term chemical N fertilizer and organic manure applications. *Glob Ecol Conserv* 22:e01011. https://doi.org/10.1016/j.gecco.2020.e01011.

de Fraiture C, Molden D, Wichelns D (2010) Investing in water for food, ecosystems, and livelihoods: An overview of the comprehensive assessment of water management in agriculture. *Agric Water Manag* 97:495–501. https://doi.org/10.1016/j.agwat.2009.08.015.

de Lara A, Longchamps L, Khosla R (2019) Soil water content and high-resolution imagery for precision irrigation: Maize yield. *Agronomy* 9:174. https://doi.org/10.3390/agronomy9040174.

El Boukhari MEM, Barakate M, Drissi BE, Bouhia Y, Lyamlouli K (2023) Seaweed Extract Biostimulants Differentially act in Mitigating Drought Stress on Faba Bean (Vicia faba L.). *J Plant Growth Regul* 42:. https://doi.org/10.1007/s00344-023-10945-w.

Elansary HO, Mahmoud EA, El-Ansary DO, Mattar MA (2020) Effects of water stress and modern biostimulants on growth and quality characteristics of mint. *Agronomy* 10:. https://doi.org/10.3390/agronomy10010006.

Ermolaev E, Jarvis Å, Sundberg C, Smårs S, Pell M, Jönsson H (2015) Nitrous oxide and methane emissions from food waste composting at different temperatures. *Waste Manag* 46:113–119. https://doi.org/10.1016/j.wasman.2015.08.021.

Fleming TR, Fleming CC, Levy CCB, Repiso C, Hennequart F, Nolasco JB, Liu F (2019) Biostimulants enhance growth and drought tolerance in Arabidopsis thaliana and exhibit chemical priming action. *Ann Appl Biol* 174:. https://doi.org/10.1111/aab.12482.

Geng SM, Yan DH, Zhang TX, Weng BS, Zhang ZB, Qin TL (2015) Effects of drought stress on agriculture soil. *Nat Hazards* 75:1997–2011. https://doi.org/10.1007/s11069-014-1409-8.

Goñi O, Quille P, O'Connell S (2018) Ascophyllum nodosum extract biostimulants and their role in enhancing tolerance to drought stress in tomato plants. *Plant Physiol Biochem* 126:63–73. https://doi.org/10.1016/j.plaphy.2018.02.024.

González Perea R, Daccache A, Rodríguez Díaz JA, Camacho Poyato E, Knox JW (2018) Modelling impacts of precision irrigation on crop yield and in-field water management. *Precis Agric* 19:497–512. https://doi.org/10.1007/s11119-017-9535-4.

Gutiérrez-Gordillo S, García-Tejero IF, García-Escalera A, Galindo P, Arco MDC, Zuazo VHD (2019) Approach to yield response of young almond trees to deficit irrigation and biostimulant applications. *Horticulturae* 5:38. https://doi.org/10.3390/horticulturae5020038.

Haghighi M, Saadat S, Abbey, Lord (2020) Effect of exogenous amino acids application on growth and nutritional value of cabbage under drought stress. *Sci Hortic (Amsterdam)* 272:109561. https://doi.org/10.1016/j.scienta.2020.109561.

He Y, Fang J, Xu W, Shi P (2022) Substantial increase of compound droughts and heatwaves in wheat growing seasons worldwide. *Int J Climatol* 42:5038–5054. https://doi.org/10.1002/joc.7518.

Hedley CB, Yule IJ (2009) Soil water status mapping and two variable-rate irrigation scenarios. In: *Precision Agriculture*. pp 342–355.

Hertel TW, Lobell DB (2014) Agricultural adaptation to climate change in rich and poor countries: Current modeling practice and potential for empirical contributions. *Energy Econ* 46:562–575. https://doi.org/10.1016/j.eneco.2014.04.014.

Hidalgo-Santiago L, Navarro-León E, López-Moreno FJ, Arjó G, González LM, Ruiz JM, Blasco B (2021) The application of the silicon-based biostimulant Codasil® offset water deficit of lettuce plants. *Sci Hortic (Amsterdam)* 285:. https://doi.org/10.1016/j.scienta.2021.110177.

Huang S, Zheng X, Luo L, Ni Y, Yao L, Ni W (2021) Biostimulants in bioconversion compost of organic waste: A novel booster in sustainable agriculture. *J. Clean. Prod.* 319.

Jaleel CA, Gopi R, Panneerselvam R (2008) Growth and photosynthetic pigments responses of two varieties of Catharanthus roseus to triadimefon treatment. *Comptes Rendus - Biol* 331:272–277. https://doi.org/10.1016/j.crvi.2008.01.004.

Jiménez-Arias D, Hernándiz AE, Morales-Sierra S, García-García AL, García-Machado FJ, Luis JC, Borges AA (2022) Applying Biostimulants to Combat Water Deficit in Crop Plants: Research and Debate. *Agronomy* 12:571. https://doi.org/10.3390/agronomy12030571.

Jithesh MN, Shukla PS, Kant P, Joshi J, Critchley AT, Prithiviraj B (2019) Physiological and Transcriptomics Analyses Reveal that Ascophyllum nodosum Extracts Induce Salinity Tolerance in Arabidopsis by Regulating the Expression of Stress Responsive Genes. *J Plant Growth Regul* 38:463–478. https://doi.org/10.1007/s00344-018-9861-4.

Khan MN, Mobin M, Abbas ZK, Alamri SA (2017) Fertilizers and their contaminants in soils, surface and groundwater. In: *Encyclopedia of the Anthropocene.* Elsevier, pp 225–240.

Łangowski Ł, Goñi O, Quille P, Stephenson P, Carmody N, Feeney E, Barton D, Østergaard L, O'Connell S (2019) A plant biostimulant from the seaweed Ascophyllum nodosum (Sealicit) reduces podshatter and yield loss in oilseed rape through modulation of IND expression. *Sci Rep* 9:. https://doi.org/10.1038/s41598-019-52958-0.

Liu X, Amjad Bashir M, Geng Y, Raza Q-U-A, Rehim A, Aon M, Luo J, Zhao Y, Zhang X, Liu H (2023) Assessment of Nutrient Leaching Losses and Crop Uptake with Organic Fertilization, Water Saving Practices and Reduced Inorganic Fertilizer. *Phyton (B Aires)* 92:1555–1570. https://doi.org/10.32604/phyton.2023.026735.

Mao XM, Zhong WW, Wang XY, Zhou XB (2017) Effects of precision planting patterns and irrigation on winter wheat yields and water productivity. *J Agric Sci* 155:1394–1406. https://doi.org/10.1017/S0021859617000508.

Mariani L, Ferrante A (2017) Agronomic management for enhancing plant tolerance to abiotic stresses—drought, salinity, hypoxia, and lodging. *Horticulturae* 3.

Mateo-Sagasta J, Zadeh SM, Turral H, Burke J (2017) Water Pollution from Agriculture: A Global Review. *Food Agric Organ* 1–35.

Melo GB, Silva AG da, Costa AC da, Silva AA da, Rosa M, Bessa LA, Rodrigues CR, Castoldi G, Vitorino LC (2024) Foliar Application of Biostimulant Mitigates Water Stress Effects on Soybean. *Agronomy* 14:414. https://doi.org/10.3390/AGRONOMY14030414.

Melo P, Abreu C, Bahcevandziev K, Araujo G, Pereira L (2020) Biostimulant effect of marine macroalgae bioextract on pepper grown in greenhouse. *Appl Sci* 10:. https://doi.org/10.3390/app10114052.

Mitra B, Chowdhury AR, Dey P, Hazra KK, Sinha AK, Hossain A, Meena RS (2022) Use of Agrochemicals in Agriculture: Alarming Issues and Solutions. In: *Input Use Efficiency for Food and Environmental Security*. Wiley, pp 85–122.

Ngoroyemoto N, Kulkarni MG, Stirk WA, Gupta S, Finnie JF, van Staden J (2020) Interactions Between Microorganisms and a Seaweed-Derived Biostimulant on the Growth and Biochemical Composition of Amaranthus hybridus L. *Nat Prod Commun* 15:. https://doi.org/10.1177/1934578X20934228.

Pahalvi HN, Rafiya L, Rashid S, Nisar B, Kamili AN (2021) Chemical Fertilizers and Their Impact on Soil Health. In: *Microbiota and Biofertilizers, Vol 2.* Springer, Cham, pp 1–20.

Panfili I, Bartucca ML, Del Buono D (2019) The treatment of duckweed with a plant biostimulant or a safener improves the plant capacity to clean water polluted by terbuthylazine. *Sci Total Environ* 646:832–840. https://doi.org/10.1016/j.scitotenv.2018.07.356.

Papantzikos V, Papanikou A, Stournaras V, Mpeza P, Mantzoukas S, Patakioutas G (2024) Biostimulant Effect of Commercial Rhizobacteria Formulation on the Growth of Vitis vinifera L.: Case of Optimal and Water Deficit Conditions. *Appl Biosci* 3:151–164. https://doi.org/10.3390/APPLBIOSCI3010010.

Paradiković N, Vinković T, Vinković Vrček I, Žuntar I, Bojić M, Medić-Šarić M (2011) Effect of natural biostimulants on yield and nutritional quality: An example of sweet yellow pepper (Capsicum annuum L.) plants. *J Sci Food Agric* 91:2146–2152. https://doi.org/10.1002/jsfa.4431.

Passioura J (2007) The drought environment: Physical, biological and agricultural perspectives. In: *Journal of Experimental Botany.* Oxford Academic, pp 113–117.

Petropoulos SA, Fernandes Â, Plexida S, Chrysargyris A, Tzortzakis N, Barreira JCM, Barros L, Ferreira ICFR (2020) Biostimulants application alleviates water stress effects on yield and chemical composition of greenhouse green bean (phaseolus vulgaris l.). *Agronomy* 10:. https://doi.org/10.3390/agronomy10020181.

Pourghasemian N, Moradi R, Naghizadeh M, Landberg T (2020) Mitigating drought stress in sesame by foliar application of salicylic acid, beeswax waste and licorice extract. *Agric Water Manag* 231:105997. https://doi.org/10.1016/j.agwat.2019.105997.

Raza Q-U-A, Bashir MA, Rehim A, Ejaz R, Raza HMA, Shahzad U, Ahmed F, Geng Y (2022) Biostimulants induce positive changes in the radish morpho-physiology and yield. *Front Plant Sci* 0:2475. https://doi.org/10.3389/FPLS.2022.950393.

Raza Q, Bashir MA, Rehim A, Geng Y, Raza HMA, Hussain S, Ahmad I, Wasif M (2023) Identifying the Role of Biostimulants in Turnip (Brassica rapa L.) Production Compared with Chemical Fertilization. *Sustain* 15:11851. https://doi.org/10.3390/su151511851.

Reddy AR, Chaitanya KV, Vivekanandan M (2004) Drought-induced responses of photosynthesis and antioxidant metabolism in higher plants. *J. Plant Physiol.* 161:1189–1202.

Rehim A, Amjad Bashir M, Raza Q-U-A, Gallagher K, Berlyn GP (2021) Yield Enhancement of Biostimulants, Vitamin B12, and CoQ10 Compared to Inorganic Fertilizer in Radish. *Agronomy* 11:697. https://doi.org/10.3390/agronomy11040697.

Rehim A, Berlyn GP, Raza QUA, Gallagher K, Raza HMA, Bashir MA (2024) Comparison of inorganic fertilizer with biostimulants and coenzyme Q10 to enhance radish performance. *Int J Veg Sci* 1–12. https://doi.org/10.1080/19315260.2024.2308265.

Rotiroti M, Sacchi E, Caschetto M, Zanotti C, Fumagalli L, Biasibetti M, Bonomi T, Leoni B (2023) Groundwater and surface water nitrate pollution in an intensively irrigated system: Sources, dynamics and adaptation to climate change. *J Hydrol* 623:129868. https://doi.org/10.1016/j.jhydrol.2023.129868.

Santos RKA, Cairo PAR, Barbosa RP, Lacerda J de J, Neto C da SM, Macedo TH de J (2019) Physiological responses of eucalyptus urophylla young plants treated with biostimulant under water deficit. *Cienc Florest* 29:. https://doi.org/10.5902/1980509826206.

Santoyo G, Guzmán-Guzmán P, Parra-Cota FI, de los Santos-Villalobos S, Orozco-Mosqueda MDC, Glick BR (2021) Plant growth stimulation by microbial consortia. *Agronomy* 11.

Souza SA, Rodrigues LN, da Cunha FF (2023) Assessing the precision irrigation potential for increasing crop yield and water savings through simulation. *Precis Agric* 24:533–559. https://doi.org/10.1007/s11119-022-09958-4.

Sujata, Goyal V, Baliyan V, Avtar R, Mehrotra S (2023) Correction to: Alleviating Drought Stress in Brassica juncea (L.) Czern and Coss. by Foliar Application of Biostimulants—Orthosilicic Acid and Seaweed Extract (Applied Biochemistry and Biotechnology, (2023), 195, 1, (693-721), 10.1007/s12010-022-04085-2). *Appl. Biochem. Biotechnol.* 195:722.

Toscano Adamo ML, Yonny ME, Villalba GF, Nazareno MA (2024) Natural biostimulants foliar application as sustainable mitigation strategy of drought stress damage on the melon crop (Cucumis melo L.). *Sci Hortic (Amsterdam)* 323:112471. https://doi.org/10.1016/j.scienta.2023.112471.

Toscano S, Gómez-Bellot MJ, Romano D, Sánchez-Blanco MJ (2023) Physiological and biochemical changes in response to Moringa oleifera biostimulant in petunia plants under water deficit. *Sci Hortic (Amsterdam)* 319:. https://doi.org/10.1016/j.scienta.2023.112187.

Touil S, Richa A, Fizir M, Argente García JE, Skarmeta Gómez AF (2022) A review on smart irrigation management strategies and their effect on water savings and crop yield. *Irrig. Drain.* 71:1396–1416.

Tyagi J, Ahmad S, Malik M (2022) Nitrogenous fertilizers: impact on environment sustainability, mitigation strategies, and challenges. *Int. J. Environ. Sci. Technol.* 19:11649–11672.

Wan LJ, Tian Y, He M, Zheng YQ, Lyu Q, Xie RJ, Ma YY, Deng L, Yi SL (2021) Effects of chemical fertilizer combined with organic fertilizer application on soil properties, citrus growth physiology, and yield. *Agric* 11:1207. https://doi.org/10.3390/agriculture11121207.

Wang X, Yang Y, Zhao J, Nie J, Zang H, Zeng Z, Olesen JE (2020) Yield benefits from replacing chemical fertilizers with manure under water deficient conditions of the winter wheat – summer maize system in the North China Plain. *Eur J Agron* 119:126118. https://doi.org/10.1016/j.eja.2020.126118.

Wang Y, Zhu Y, Zhang S, Wang Y (2018) What could promote farmers to replace chemical fertilizers with organic fertilizers? *J Clean Prod* 199:882–890. https://doi.org/10.1016/j.jclepro.2018.07.222.

Ward FA (2022) Enhancing climate resilience of irrigated agriculture: A review. *J. Environ. Manage.* 302:114032.

White D a. (2010) Quantification of Agricultural Drought for Effective Drought Mitigation and Preparedness: Key Issues and Challenges. *Agric Drought Indices Proc a WMO Expert Meet* held Murcia, Spain.

Zaheer IE, Ali S, Saleem MH, Yousaf HS, Malik A, Abbas Z, Rizwan M, Abualreesh MH, Alatawi A, Wang X (2022) Combined application of zinc and iron-lysine and its effects on morpho-physiological traits, antioxidant capacity and chromium uptake in rapeseed (Brassica napus L.). *PLoS One* 17:e0262140. https://doi.org/10.1371/journal.pone.0262140.

Zhang X, Bashir MA, Raza QUA, Liu X, Luo J, Zhao Y, Lei Q, Raza HMA, Rehim A, Geng Y, Liu H (2023) Evaluating the effects of sustainable chemical and organic fertilizers with water saving practice on corn production and soil characteristics. *Phyton-International J Exp Bot* 92:1349–1360. https://doi.org/10.32604/phyton.2023.026952.

Biographical Sketches

Qurat-Ul-Ain Raza

Affiliation: Department of Biological and Environmental Sciences, Emerson University Multan, 60000, Pakistan

Department of Soil Science, Faculty of Agricultural Sciences and Technology, Bahauddin Zakariya University, Multan, 60800, Pakistan

Education: Ph.D. Soil Science

Research and Professional Experience: Research work related to the application of biostimulants in agricultural sustainability to minimize the use of chemical fertilizers. I also worked for waste management in agriculture specifically sugar industrial waste products.

Publications from the Last 3 Years:

[1] Raza, Qurat-Ul-Ain, Abdur Rehim, Muhammad Amjad Bashir, Hafiz Muhammad Ali Raza, Muhammad Aon, Yucong Geng, Mahmoud Moustafa, Mohammed O. Alshaharni, Haider Ali, and Rosa Sanchez Lucas. "Identifying the abilities of biostimulants to improve vegetable production compared with conventional fertilizer." *Frontiers in Sustainable Food Systems* 8 (2024): 1345259.

[2] Raza, Qurat-Ul-Ain, Muhammad Amjad Bashir, Abdur Rehim, Yucong Geng, Hafiz Muhammad Ali Raza, Sajjad Hussain, Ijaz Ahmad, and Muhammad Wasif. "Identifying the role of biostimulants in turnip (Brassica rapa L.) production compared with chemical fertilization." *Sustainability* 15, no. 15 (2023): 11851.

[3] Raza, Qurat-Ul-Ain, Muhammad Amjad Bashir, Abdur Rehim, Umbreen Shahzad, and Yucong Geng. "Biostimulants induce positive changes in the radish morpho-physiology and yield." *Frontiers in Plant Science* 13 (2022): 950393.

Abdur Rehim

Affiliation: Department of Soil Science, Faculty of Agricultural Sciences and Technology, Bahauddin Zakariya University, Multan, 60800, Pakistan
Education: Ph.D. Soil Science, Post Doc (Yale, University)
Research and Professional Experience: Research work in Soil Fertility and Plant Nutrition
Professional Appointments: Working as Associate Professor in Department of Soil Science, Faculty of Agricultural Sciences and Technology, Bahauddin Zakariya University, Multan, 60800, Pakistan

Publications from the Last 3 Years:

[1] Rehim, Abdur, Graeme P. Berlyn, Qurat-Ul-Ain Raza, Kevin Gallagher, Hafiz Muhammad Ali Raza, and Muhammad Amjad Bashir. "Comparison of inorganic fertilizer with biostimulants and coenzyme Q10 to enhance radish performance." *International Journal of Vegetable Science* (2024): 1-12.
[2] Aon, Muhammad, Zeshan Aslam, Shahid Hussain, Muhammad Amjad Bashir, Muhammad Shaaban, Sajid Masood, Sidra Iqbal et al. "Wheat Straw biochar produced at a low temperature enhanced maize growth and yield by influencing soil properties of typic calciargid." *Sustainability* 15, no. 12 (2023): 9488.
[3] Liu, Xiao-tong, Muhammad Amjad Bashir, Yu-cong Geng, Qurat-Ul-Ain Raza, Abdur Rehim, Muhammad Aon, Jian-hang Luo, Ying Zhao, Xue-jun Zhang, and Hong-bin Liu. "Assessment of nutrient leaching losses and crop uptake with organic fertilization, water saving practices and reduced inorganic fertilizer." *Phyton-International J Exp Bot* (2023): 1555-1570.

Asad Ismail Noor

Affiliation: Department of Soil Science, Faculty of Agricultural Sciences and Technology, Bahauddin Zakariya University, Multan, 60800, Pakistan
Education: Bachelor's in Agriculture Soil Science, Currently Enrolled in Master's in Soil Science.
Research and Professional Experience: Research work in biostimulants performance as fertilizer reduction technique.

Publications from the Last 3 Years:

[1] Raza, Qurat-ul-Ain, Muhammad Amjad Bashir, Abdur Rehim, Faiz Ul Hassan, Hafiz Muhammad Ali Raza, Asad Ismail Noor, and Muhammad Usama Iqbal. "Melatonin: A Promising Tool Against Climate Change and Food Security for Better Plant

Survival." In *Melatonin in Plants: A Regulator for Plant Growth and Development*, pp. 291-300. Singapore: Springer Nature Singapore, 2023.

Ijaz Ahmad

Affiliation: Department of Agronomy, Engro Fertilizers Ltd. Lahore, 54600, Pakistan

Education: M.Sc. Soil Science

Research and Professional Experience: Working as Deputy Manager (Regulatory Affairs) at Engro Fertilizers Ltd. Lahore, Pakistan. Research interests related to agriculture productivity and soil health.

Publications from the Last 3 Years:

[1] Ali MA, Naeem F, Tariq N, Ahmed I and Imran A (2021) Bioactive Nutrient Fortified Fertilizer: A Novel Hybrid Approach for the Enrichment of Wheat Grains With Zinc. *Front. Plant Sci.* 12:743378. doi: 10.3389/fpls.2021.743378

[2] Shah SSH, Azhar M, Nadeem F, Ali MA, Khan MN, Ahmad I, et al. (2023) Enhancements in yield, agronomic, and zinc recovery efficiencies of rice-wheat system through bioactive zinc coated urea application in Aridisols. *PLoS ONE* 18(3): e0282615. https://doi.org/10.1371/journal.pone.0282615

[3] Raza, Qurat-Ul-Ain, Abdur Rehim, Muhammad Amjad Bashir, Hafiz Muhammad Ali Raza, Muhammad Aon, Yucong Geng, Mahmoud Moustafa, Mohammed O. Alshaharni, Haider Ali, and Rosa Sanchez Lucas. "Identifying the abilities of biostimulants to improve vegetable production compared with conventional fertilizer." *Frontiers in Sustainable Food Systems* 8 (2024): 1345259.

Faiz Ul Hassan

Affiliation: Department of Soil Science, College of Agriculture, University of Layyah, Layyah, 31200, Pakistan

Education: Bachelor's in Agriculture Soil Science.

Research and Professional Experience: Research work related to biostimulants application in agriculture sustainability.

Publications from the Last 3 Years:

[1] Raza, Qurat-ul-Ain, Muhammad Amjad Bashir, Abdur Rehim, Faiz Ul Hassan, Hafiz Muhammad Ali Raza, Asad Ismail Noor, and Muhammad Usama Iqbal. "Melatonin: A Promising Tool Against Climate Change and Food Security for Better Plant

Survival." In *Melatonin in Plants: A Regulator for Plant Growth and Development*, pp. 291-300. Singapore: Springer Nature Singapore, 2023.

Muhammad Amjad Bashir

Affiliation: Sichuan Provincial Key Laboratory of Philosophy and Social Sciences for Monitoring and Evaluation of Rural Land Utilization, Chengdu Normal University, Chengdu 611130, P. R. China

Education: Ph.D. Agroecology

Research and Professional Experience: Research work related to the rice-co-culture systems, biostimulants in agricultural sustainability, improvement of soil health, and controlling greenhouse gas emissions for environmental sustainability.

Publications from the Last 3 Years:

[1] Li, Weijing, Yibin Zhang, Hongyuan Wang, Bingqian Fan, Muhammad Amjad Bashir, Ke Jin, and Hongbin Liu. "Legacy effect of long-term rice–crab co-culture on N2O emissions in paddy soils." *Applied Soil Ecology* 196 (2024): 105251.

[2] Dai, Fuyue, Bingqian Fan, Jungai Li, Yitao Zhang, Hongyuan Wang, Zhen Wang, Muhammad Amjad Bashir et al. "Fate of 15N-labelled urea as affected by long-term manure substitution." *Science of The Total Environment* 893 (2023): 164924.

[3] Rehim, Abdur, Graeme P. Berlyn, Qurat-Ul-Ain Raza, Kevin Gallagher, Hafiz Muhammad Ali Raza, and Muhammad Amjad Bashir. "Comparison of inorganic fertilizer with biostimulants and coenzyme Q10 to enhance radish performance." *International Journal of Vegetable Science* (2024): 1-12.

[4] Aon, Muhammad, Zeshan Aslam, Shahid Hussain, Muhammad Amjad Bashir, Muhammad Shaaban, Sajid Masood, Sidra Iqbal et al. "Wheat Straw biochar produced at a low temperature enhanced maize growth and yield by influencing soil properties of typic calciargid." *Sustainability* 15, no. 12 (2023): 9488.

Epilogue

As we conclude the final chapter of *Irrigation Management: Strategies, Sustainability, and Agricultural Advancements*, one central theme resonates: the future of agriculture hinges on our water management. The knowledge, technologies, and practices presented here are not merely tools but essential lifelines for addressing global challenges such as food security, climate resilience, and sustainable development.

This book has shown that irrigation is far more than the mechanical delivery of water - it is a nexus of technology, science, and policy. From drip systems that lower water losses to artificial intelligence solutions that adjust resource allocation, we stand on the edge of an agricultural revolution. However, for these solutions to reach their full capability, cooperation is crucial. Governments must implement policies that support sustainable practices, researchers must continue to innovate, and farmers must have access to the necessary knowledge and tools.

The stories and case studies within these pages are a testament to human determination. We have observed how communities in arid areas leverage technology to overcome drought issues, how precision agriculture converts irrigation into an art of efficiency, and how worldwide collaboration can yield strategies that transcend borders. Each success story is a beacon of what can be achieved when ingenuity meets resolve.

Yet, challenges continue. Climate change remains to strain water resources, and inequalities in access to technology and infrastructure threaten to widen the gap between those who can and cannot adjust. As we move forward, it is imperative that solutions are inclusive, equitable, and adaptable to the diverse realities of farmers worldwide.

Looking ahead, the call to action is clear: we must endow more in research and development strategies and work, empowering education and capacity-building programs, and incorporate sustainability into every aspect of water management. By doing so, we can ensure that future generations inherit a world where agriculture thrives in harmony with the environment. We hope that the knowledge in these pages inspire continued innovation, foster dialogue among stakeholders, and empower action on the ground. Together, we can chart a course toward a resilient agricultural system-one that sustains not only crops, but also the communities and ecosystems that depend on them.

Index

A

B

C

D

E

F

G

H

I

K

L

M

N

W